Aquariums and zoos in Tokyo where penguins can be seen and the number of each kind

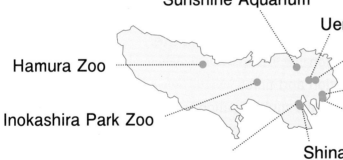

Sunshine Aquarium

Ueno Zoological Gardens

Sumida Aquarium

Hamura Zoo

Edogawa City Shizen Zoo

Tokyo Sea Life Park

Inokashira Park Zoo

Shinagawa Aquarium

Maxell Aqua Park Shinagawa

01204

Location	Penguin kind	Count
Maxell Aqua Park Shinagawa (February, 2018)	King Penguin	3
	Northern Rockhopper Penguin	7
	African Penguin	27
	Gentoo Penguin	6
Inokashira Park Zoo (November, 2018)	Humboldt Penguin	4
Ueno Zoological Gardens (November, 2018)	African Penguin	29
Tokyo Sea Life Park (November, 2018)	King Penguin	8
	Little Penguin	41
	Humboldt Penguin	120
	Southern Rockhopper Penguin	42
Sunshine Aquarium (February, 2018)	African Penguin	60
Shinagawa Aquarium (December, 2018)	Magellanic Penguin	15
Edogawa City Shizen Zoo (February, 2018)	Humboldt Penguin	10
Sumida Aquarium (November, 2018)	Magellanic Penguin	55
Hamura Zoo (December, 2018)	Humboldt Penguin	19

JN119578

※ Those penguins may not be seen according to seasons or their conditions.

Table of Contents

Let's learn mathematics together.

Yui

Hiroto

Nanami

Daiki

Important words and rules

Rules that you found

Let's **deepen.**

You will want to learn much more.

Want to connect

Solve new problems.

Let's explore the number of children riding in
the amusement rides.

Let's think about the meaning of multiplication and how to calculate.

The children came to the amusement park for their field trip.

1 Let's explore the number of children riding in the amusement rides.

① How many children are riding in the ⊖ (Ferris wheel) altogether?

② How many children are riding in the ☕ (coffee cup) altogether?

③ What is the difference between the numbers of children in each ⊖ and each ☕?

The total number of children in the ☕ can be represented as follows: 3 children in each cup, for 6 cups, there are 18 children.

> It's easy to represent the total number if the same number of children are in every cup.

④ Let's also represent the total number of children riding in other rides in the same way.

2 children in each ✈ (plane), for 4 planes, there are 8 children.

☐ children in each 🏎 (Go-cart), for ☐ carts, there are ☐ children.

☐ children in each 🎢 (roller coaster), for ☐ coasters, there are ☐ children.

The total number of children riding in the Go-carts is represented as follows:

4 children in each cart, for **3 carts**, there are **12 children**.

This is written as $4 \times 3 = 12$, and is read as "4 multiplied by 3 equals 12."*

$$4 \quad \times \quad 3 \quad = \quad 12$$

4	3	12
Number of children in each cart	Number of carts	Total number

This kind of operation is called **multiplication**.

Want to represent

1 ▶ Let's look at the picture in pages 5 and 6 and write math sentences which represent the total number of children riding in the following rides.

Number of children in each ride	Number of rides	Total number

$3 \times 6 = \boxed{}$

$\boxed{} \times \boxed{} = \boxed{}$

train

$\boxed{} \times \boxed{} = \boxed{}$

$\boxed{} \times \boxed{} = \boxed{}$

Thinking of the number of children riding in each ride.

Nanami

Multiplication is the operation that is used to get a total number when there is the same number in each set.

*Japanese textbook uses Japanese notation. In English notation, it should be $3 \times 4 = 12$. In Japanese, the verb comes to the last in the sentence.

2 There are the same number of objects in each set. Let's represent them in multiplication sentences.

① **3** boxes of chocolates

② **6** packs of fish

③ **5** bags of jellies

④ **2** plates of pears

Want to try

2 How many kiwi fruits are there in total? Let's examine the number by using blocks and represent it in a multiplication sentence.

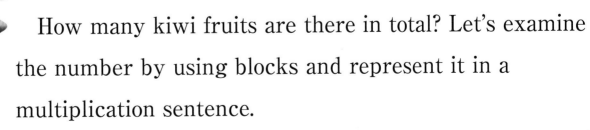

3 There are 8 cans in each box. Let's examine how many cans there are in 6 boxes.

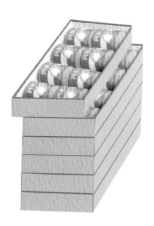

① Let's represent it in a multiplication sentence.

| Number of cans in each box | Number of boxes | Total number |

② How many cans are there in total?

The answer to 8 × 6 is the same as the answer to 8 + 8 + 8 + 8 + 8 + 8.

Way to see and think

You can find the answer easily by using blocks.

3 Let's represent the following problems in multiplication expressions and find the total number.

① There are 4 cakes in each box. How many cakes are there in 4 boxes?

② There are 6 wedges of cheese in each box. How many wedges are there in 7 boxes?

4 There are 12 lemons. Let's put the same number into some bags. In what ways can you put lemons into the bags? Let's represent them in multiplication sentences.

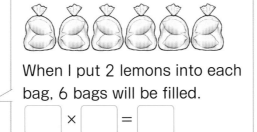

When I put 2 lemons into each bag, 6 bags will be filled.

[ ] × [ ] = [ ]

Yui

How about putting 3 lemons into each bag?

Nanami

There seem to be other ways.

Daiki

4 Let's look for some situations around us that can be represented in multiplication sentences.

Multiplication Hunting

Number of cherries
2 × 6 = 12

Become a writing master.

Summary notebook

Let's summarize what you learned that day.

Write today's date.

October 6

Write the problem of the day.

> Let's explore the number of children riding in the amusement rides.

Coffee cup : 3 children in each cup, for 6 cups, there are 18 children.

Don't erase your mistakes.

Plane : 2 children in each plane, for 4 planes, there are ~~6~~ children.
                                    8

Go-cart : 4 children in each cart, for 3 carts, there are 12 children.

Roller coaster : 6 children in each coaster, for 3 coasters, there are 18 children.

Write what you newly learned.

> The total number of children riding in the roller coaster is written as
>
> $$6 \times 3 = 18.$$
>
> This kind of operation is called multiplication.

It's easy to see by using colors.

> The total number of children riding in the roller coaster is written as
>
> $$6 \times 3 = 18.$$
>
> This kind of operation is called multiplication.

Let's write math sentences which represent the total number of children riding in the other rides.

Number of children   Number of   Total
in each ride   rides   number

$$3 \times 6 = 18$$
$$2 \times 4 = 8$$
$$6 \times 3 = 18$$
$$5 \times 4 = 20$$

Coffee cup $3 \times 6 = 18$
Plane $2 \times 4 = 8$
Go-cart $4 \times 3 = 12$
Train $5 \times 4 = 20$

Summary

Multiplication is the operation that is used to get a total number when there is the same number in each set.

< Reflection >

○ When there is the same number in each set, it can be represented in a multiplication sentence.

○ If that can be represented in a math sentence, can I calculate it as an addition and a subtraction?

○ I want to represent it in different multiplication sentences.

As for reflection, the following must be written:
· what you understood
· what you noticed
· what you definitely achieved
· what you didn't understand
· what you want to do more

Want to know

1 There are some pieces of 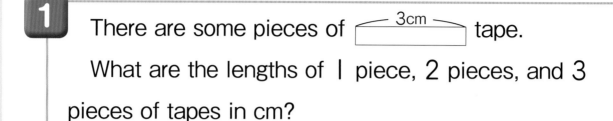 tape.

What are the lengths of 1 piece, 2 pieces, and 3 pieces of tapes in cm?

1 piece

2 pieces

3 pieces

1 piece	3 × 1 = 3	☐ cm
2 pieces	3 × ☐ = ☐	☐ cm
3 pieces	3 × ☐ = ☐	☐ cm

The terms 1 piece, 2 pieces, and 3 pieces are also called 1 **time**, 2 times, and 3 times.*

Want to confirm

 There are 4 books stacked together and each book is 2 cm thick. How many times of the thickness of one book is the total height of the books? What is the height in cm?

*Times is 'bai' in Japanese. It is used on the context such as '2 times of 3 cm.' In that sense, 'cm' means the measurement unit.

3 Row of 5 in the multiplication table

Want to explore

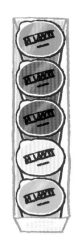

1 There are 5 jellies in each box. Let's examine the total number of jellies.

(1) Let's examine the total number of jellies as the number of boxes increases from 1 to 4.

Let's think by using blocks.

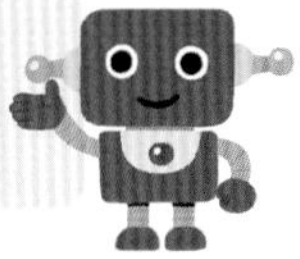

Number of jellies in each box / Number of boxes / Total number

For 1 box

$5 \times 1 = \square$

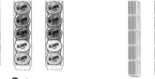

For 2 boxes

$5 \times 2 = \square$

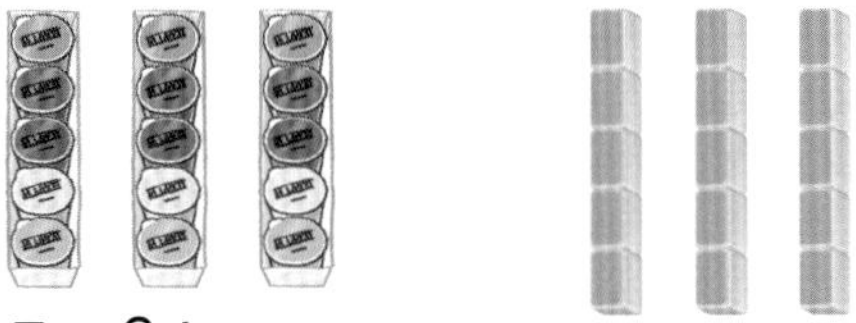

For 3 boxes

$5 \times 3 = \square$

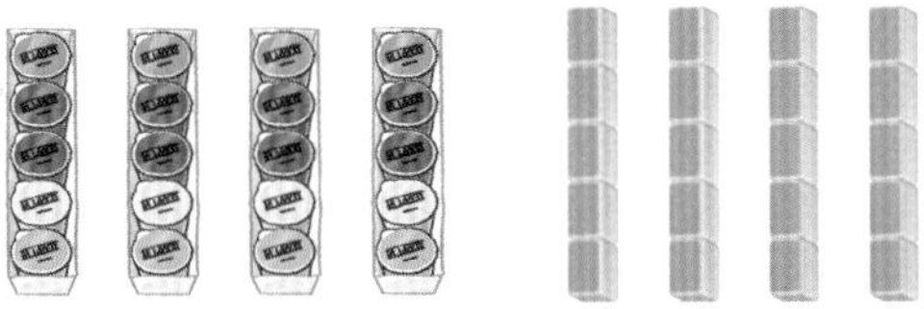

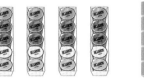

For 4 boxes

$5 \times 4 = \square$

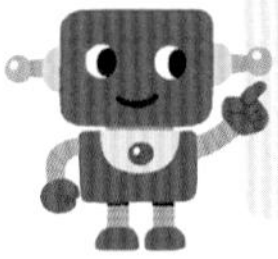

When the number of boxes increases by 1, by how many does the total number of jellies increase?

② Let's examine the total number as the number of boxes increases from 5 to 9.

For 5 boxes

$5 \times 5 = \boxed{}$

For 6 boxes

$5 \times 6 = \boxed{}$

For 7 boxes

$5 \times 7 = \boxed{}$

For 8 boxes

$5 \times 8 = \boxed{}$

For 9 boxes

$5 \times 9 = \boxed{}$

$5 \times 1 = 5$, $5 \times 2 = 10$, and $5 \times 3 = 15$ can be read as "five ones is five, five twos is ten, and five threes is fifteen." *

This way of reading multiplication sentences as a set of the same number such as 5 is called row of 5 in the **multiplication table**, "kuku" in Japanese.

*In English, 5×3 is read and memorized as '5 times 3.' In Japanese, it means '5, 3 times' or '3 times of 5.' 'Five threes' is used for memorization in Japan.

2 Let's memorize the row of 5 in the multiplication table.

Row of 5

5 × 1 = 5	five ones is five
5 × 2 = 10	five twos is ten
5 × 3 = 15	five threes is fifteen
5 × 4 = 20	five fours is twenty
5 × 5 = 25	five fives is twenty-five
5 × 6 = 30	five sixes is thirty
5 × 7 = 35	five sevens is thirty-five
5 × 8 = 40	five eights is forty
5 × 9 = 45	five nines is forty-five

By how many does the answer in the row of 5 increase?

1 Let's make multiplication cards for the row of 5 and practice with them.

front

back

5 × 6   30

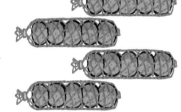

2 There are 5 oranges in each bag. How many oranges are there in 4 bags?

Math Sentence : [   ]   Answer : [   ] oranges

3 There are 7 children and each child is given 5 sheets of origami paper.

How many sheets are needed altogether?

Math Sentence : [   ]   Answer : [   ] sheets

Want to explore

1 There are 2 cakes on each plate.

Let's examine the total number of cakes.

① Let's examine the total number of cakes as the number of plates increases from 1 to 4.

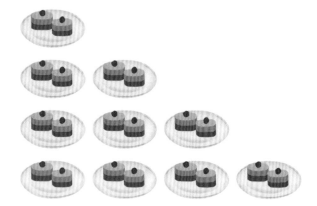

$2 \times 1 = \boxed{}$

$2 \times 2 = \boxed{}$

$2 \times 3 = \boxed{}$

$2 \times 4 = \boxed{}$

$2 \times 5 = \boxed{}$

$2 \times 6 = \boxed{}$

$2 \times 7 = \boxed{}$

$2 \times 8 = \boxed{}$

$2 \times 9 = \boxed{}$

② Let's examine the total number as the number of plates increases from 5 to 9.

③ When the number of plates increases by 1, by how many does the total number of cakes increase?

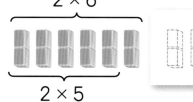

2×6

2×5

2 Let's memorize the row of 2 in the multiplication table.

By how many does the answer in the row of 2 increase?

1 Let's make multiplication cards for the row of 2 and practice with them.

front
2×5

back
10

Row of 2

$2 \times 1 = 2$	two ones is two
$2 \times 2 = 4$	two twos is four
$2 \times 3 = 6$	two threes is six
$2 \times 4 = 8$	two fours is eight
$2 \times 5 = 10$	two fives is ten
$2 \times 6 = 12$	two sixes is twelve
$2 \times 7 = 14$	two sevens is fourteen
$2 \times 8 = 16$	two eights is sixteen
$2 \times 9 = 18$	two nines is eighteen

2 2 people are sitting on each bench. There are 6 benches. How many people are there altogether?

Math Sentence : [   ]   Answer : [   ] people

3 Each child makes 2 origami cranes. How many cranes can be made by 7 children?

Math Sentence : [   ]   Answer : [   ] cranes

1 Let's examine the total number of wheels of tricycles.

① Let's examine the total number of wheels as the number of tricycles increases from 1 to 4.

$3 \times 1 =$ ☐

$3 \times 2 =$ ☐

$3 \times 3 =$ ☐

$3 \times 4 =$ ☐

$3 \times 5 =$ ☐

$3 \times 6 =$ ☐

$3 \times 7 =$ ☐

$3 \times 8 =$ ☐

$3 \times 9 =$ ☐

② Let's examine the total number as the number of tricycles increases from 5 to 9.

③ When the number of tricycles increases by 1, by how many does the total number of wheels increase?

3×4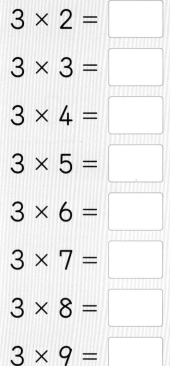

3×5

Row of 3

2 Let's memorize the row of **3** in the multiplication table.

3 × 1 = 3	three ones is three
3 × 2 = 6	three twos is six
3 × 3 = 9	three threes is nine
3 × 4 = 12	three fours is twelve
3 × 5 = 15	three fives is fifteen
3 × 6 = 18	three sixes is eighteen
3 × 7 = 21	three sevens is twenty-one
3 × 8 = 24	three eights is twenty-four
3 × 9 = 27	three nines is twenty-seven

By how many does the answer in the row of 3 increase?

1 Let's make multiplication cards for the row of **3** and practice with them.

front
back

$$3 \times 2 \qquad 6$$

2 You bought **8** candies that cost **3** yen each. How much was the total cost?

Math Sentence : [   ]   Answer : [   ] yen

3 If you read **3** pages of a book every day, how many pages can you read in **5** days?

Math Sentence : [   ]   Answer : [   ] pages

21

6 Row of 4 in the multiplication table

Want to explore

1 There are cars that have 4 wheels each. Let's examine the total number of wheels.

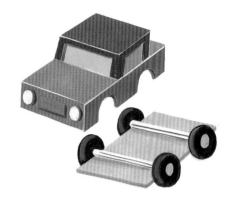

① Let's examine the total number of wheels as the number of cars increases from 1 to 4.

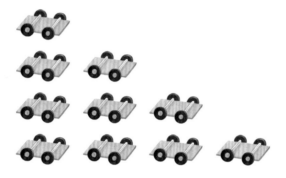

$4 \times 1 = \square$

$4 \times 2 = \square$

$4 \times 3 = \square$

$4 \times 4 = \square$

$4 \times 5 = \square$

$4 \times 6 = \square$

$4 \times 7 = \square$

$4 \times 8 = \square$

$4 \times 9 = \square$

② Let's examine the total number as the number of cars increases from 5 to 9.

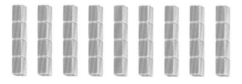

③ When the multiplier 5 increases by 1 from 4 × 5 to 4 × 6, by how many does the answer increase?

Multiplicand　Multiplier　Answer

$4 \times \boxed{5} = 20$

increase by 1 ↓　　↓ increase by $\square$

$4 \times \boxed{6} = \square$

Row of 4

4 × 1 = 4	four ones is four
4 × 2 = 8	four twos is eight
4 × 3 = 12	four threes is twelve
4 × 4 = 16	four fours is sixteen
4 × 5 = 20	four fives is twenty
4 × 6 = 24	four sixes is twenty-four
4 × 7 = 28	four sevens is twenty-eight
4 × 8 = 32	four eights is thirty-two
4 × 9 = 36	four nines is thirty-six

2 Let's memorize the row of 4 in the multiplication table.

By how many does the answer in the row of 4 increase?

1 Let's make multiplication cards for the row of 4 and practice with them.

front
4 × 8

back
32

2 Class 3 in the second grade can be divided into 6 groups of 4 children. How many children are there altogether?

Math Sentence : _____   Answer : ____ children

3 There are 4 dL of juice in each bottle. What is the total amount of juice in dL for 3 bottles?

Math Sentence : _____   Answer : ____ dL

Row of 2	Row of 3	Row of 4	Row of 5
2 × 1 = 2	3 × 1 = 3	4 × 1 = 4	5 × 1 = 5
2 × 2 = 4	3 × 2 = 6	4 × 2 = 8	5 × 2 = 10
2 × 3 = 6	3 × 3 = 9	4 × 3 = 12	5 × 3 = 15
2 × 4 = 8	3 × 4 = 12	4 × 4 = 16	5 × 4 = 20
2 × 5 = 10	3 × 5 = 15	4 × 5 = 20	5 × 5 = 25
2 × 6 = 12	3 × 6 = 18	4 × 6 = 24	5 × 6 = 30
2 × 7 = 14	3 × 7 = 21	4 × 7 = 28	5 × 7 = 35
2 × 8 = 16	3 × 8 = 24	4 × 8 = 32	5 × 8 = 40
2 × 9 = 18	3 × 9 = 27	4 × 9 = 36	5 × 9 = 45

Nanami

7 Finding the rules

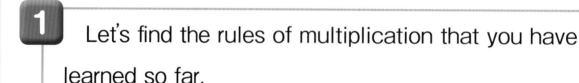

Want to discuss   Rules of multiplication

1 Let's find the rules of multiplication that you have learned so far.

① Nanami drew lines on the blackboard as shown above. Let's talk about the rules that she found.

② Let's talk about other rules that you found.

She is looking at the answers in the row of 2 and row of 3.

$$3 \times 9 \qquad 27$$

1   Let's write multiplication expressions and answers in the row of 2, 3, 4, and 5 on each card and play the card game.

①   Picking up answers

How to play

(1) Show an expression card.

(2) Pick up a card that shows the answer to the expression.

Let's play with the cards in each row, and then with all cards.

②   Comparing sizes

How to play

(1) Place expression cards face down.

(2) Turn over the card one by one and compare the sizes of the answer.

What you can do now

☐ Understanding the meaning of multiplication.

1 Let's fill in each ☐ with a number.

① The total number of apples can be represented as follows:

☐ apples in each bag, for ☐ bags, there are ☐ apples.

This is written as ☐ × ☐ = ☐

Number of apples in each bag   Number of bags   Total number

② 7 sets of 5 can be called ☐ times 5.

☐ Can do multiplications by using the multiplication table.

2 Let's calculate the following.

①	2×2	②	3×1	③	2×8	④	5×9
⑤	3×2	⑥	2×9	⑦	4×8	⑧	4×5
⑨	2×5	⑩	5×1	⑪	5×6	⑫	4×2

☐ Can make a multiplication expression and find the answer.

3 Let's answer the following.

① There are 2 pieces of sushi on each plate. There are 4 plates. How many pieces of sushi are there altogether?

② There is a 3 cm tape. What is the length that is three times the length of this tape in cm?

Supplementary Problems p.128

Usefulness and efficiency of learning

1 2 × 4 is a math expression that represents the total number of oranges.

Which situation is represented by this expression, Ⓐ or Ⓑ?

Ⓐ
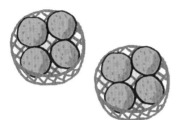

Ⓑ
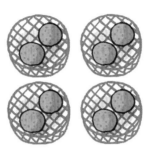

☐ Understanding the meaning of multiplication.

2 Let's calculate the following.

① 5 × 3   ② 3 × 6   ③ 5 × 8
④ 4 × 1   ⑤ 5 × 2   ⑥ 2 × 3
⑦ 3 × 9   ⑧ 4 × 4   ⑨ 4 × 7
⑩ 5 × 5   ⑪ 4 × 9   ⑫ 2 × 1

☐ Can do multiplications by using the multiplication table.

3 There are 3 eggplants in each bag. How many eggplants are there in 7 bags?

☐ Can make a multiplication expression and find the answer.

4 What multiplication expression do the blocks shown on the right represent?

Let's choose the answer from Ⓐ, Ⓑ, Ⓒ, and Ⓓ.

Ⓐ 3 × 4   Ⓑ 5 × 3
Ⓒ 2 × 6   Ⓓ 4 × 3

Let's deepen.

In what situation can we use multiplication?

Yui

27

Deepen.

Making multiplication problems

① Nanami looked at the above picture and made a multiplication problem.

Let's fill in each ☐ with a number.

I made a math problem by using apples.

Nanami

There are ☐ apples in each basket.
How many apples are there altogether?
Math expression : ☐ × ☐

② Let's make a multiplication problem as Nanami.

③ Let's make a problem for a multiplication sentence from your surroundings and write it on the card.

Let's make a presentation on the problem you made.

12 Multiplication (2)
Let's find the rules of multiplication and make the multiplication table.

Want to explore

1 There are 6 wedges of cheese in each box.

Let's examine the total number of wedges.

① Let's examine the total number of wedges as the number of boxes increases from 1.

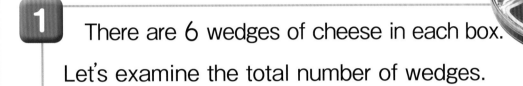

$$6 \times 1 \qquad 6 \times 2 \qquad 6 \times 3$$

② When the multiplier 4 increases by 1 from 6×4 to 6×5, by how many does the answer increase?

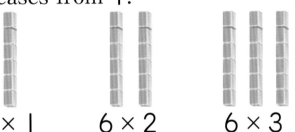

$$6 \times \boxed{4} = 24$$
$$6 \times \boxed{5} = 30$$

increase by ☐

③ Let's make the row of 6 in the multiplication table.

$6 \times 1 =$ ☐
$6 \times 2 =$ ☐
$6 \times 3 =$ ☐
$6 \times 4 =$ ☐
$6 \times 5 =$ ☐
$6 \times 6 =$ ☐
$6 \times 7 =$ ☐
$6 \times 8 =$ ☐
$6 \times 9 =$ ☐

When the multiplier increases by 1, the answer increases by the multiplicand.

2 Let's memorize the row of 6 in the multiplication table.

 1 Let's make multiplication cards for the row of 6 and practice with them.

front
back

6 × 4   24

Row of 6	
6 × 1 = 6	six ones is six
6 × 2 = 12	six twos is twelve
6 × 3 = 18	six threes is eighteen
6 × 4 = 24	six fours is twenty-four
6 × 5 = 30	six fives is thirty
6 × 6 = 36	six sixes is thirty-six
6 × 7 = 42	six sevens is forty-two
6 × 8 = 48	six eights is forty-eight
6 × 9 = 54	six nines is fifty-four

Want to confirm

 2 Let's find the following numbers by multiplication.

① Total number of goldfish   ② Total number of doughnuts

Want to discuss

3 Let's think about a math expression to find the total number for the picture on the right.

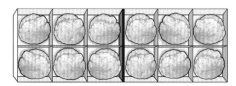

Since there are 2 boxes of 6 pieces, ...

Can I think by using the row of 2 or row of 3?

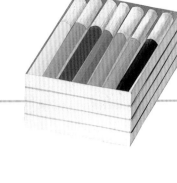

Row of 7 in the multiplication table

Want to explore

1 There are 7 pens in each box. Let's examine the total number of pens.

① Let's examine the total number of pens as the number of boxes increases from 1 and make the row of 7 in the multiplication table.

7×1 7×2 7×3

$7 \times 1 =$ ☐
$7 \times 2 =$ ☐
$7 \times 3 =$ ☐
$7 \times 4 =$ ☐
$7 \times 5 =$ ☐
$7 \times 6 =$ ☐
$7 \times 7 =$ ☐
$7 \times 8 =$ ☐
$7 \times 9 =$ ☐

When the multiplier increases by 1, ...

The answer to 7×3 plus ☐ equals the answer to 7×4.

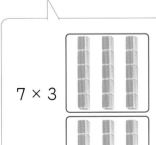

7×3 5×3 Thinking about adding 5×3 and 2×3 ...

2×3

31

2 Let's memorize the row of 7 in the multiplication table.

Row of 7		
$7 \times 1 = 7$	seven ones is seven
$7 \times 2 = 14$	seven twos is fourteen
$7 \times 3 = 21$	seven threes is twenty-one
$7 \times 4 = 28$	seven fours is twenty-eight
$7 \times 5 = 35$	seven fives is thirty-five
$7 \times 6 = 42$	seven sixes is forty-two
$7 \times 7 = 49$	seven sevens is forty-nine
$7 \times 8 = 56$	seven eights is fifty-six
$7 \times 9 = 63$	seven nines is sixty-three

 Let's make multiplication cards for the row of 7 and practice with them.

front

7×6

back

42

Want to confirm

 There are 7 days in a week. How many days are there in 4 weeks?

Sun.	Mon.	Tue.	Wed.	Thu.	Fri.	Sat.
1	2	3	4	5	6	7
8	9	10	11	12	13	14
15	16	17	18	19	20	21

Want to discuss

3 Your friend cannot find the answer to 7×6. How should you help him or her?

Let's explain to him or her by using a math expression.

Can you find the answer to 6×7?

Want to solve

1 Each child uses a tape that is 8 cm long. How many cm of tape are needed for 3 children?

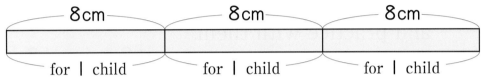

① Let's write a multiplication expression.

② Let's find the answer.

Want to improve

▶ **1** Let's find an easier way to make the row of 8 in the multiplication table.

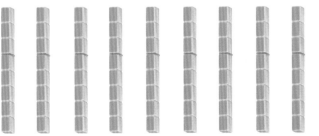

Daiki

When the multiplier increases by 1, the answer ...

As for 8 × 2, decompose 8 into 5 and 3 ...

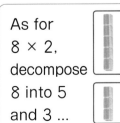

Nanami

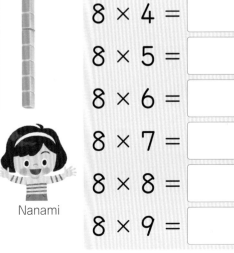

8 × 1 = ☐
8 × 2 = ☐
8 × 3 = ☐
8 × 4 = ☐
8 × 5 = ☐
8 × 6 = ☐
8 × 7 = ☐
8 × 8 = ☐
8 × 9 = ☐

2 Let's memorize the row of **8** in the multiplication table.

Row of 8		
$8 \times 1 = 8$	eight ones is eight
$8 \times 2 = 16$	eight twos is sixteen
$8 \times 3 = 24$	eight threes is twenty-four
$8 \times 4 = 32$	eight fours is thirty-two
$8 \times 5 = 40$	eight fives is forty
$8 \times 6 = 48$	eight sixes is forty-eight
$8 \times 7 = 56$	eight sevens is fifty-six
$8 \times 8 = 64$	eight eights is sixty-four
$8 \times 9 = 72$	eight nines is seventy-two

2 ▶ Let's make multiplication cards for the row of **8** and practice with them.

front
back

8×4 32

3 ▶ Each child receives **8** sheets of colored paper.

How many sheets are needed for **6** children?

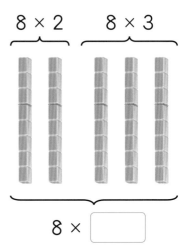

4 ▶ What is the sum of the answers to **8 × 2** and **8 × 3**?

By what number should we multiply **8** to get the same answer as the sum?

8×2 8×3

$8 \times \boxed{}$

34

4 Row of 9 in the multiplication table

Want to solve

1 Each baseball team has 9 players.

How many players are there in 4 teams?

① Let's make a multiplication expression.

② Let's find the answer.

Want to improve

 1 Let's find an easier way to make the row of 9 in the multiplication table.

$9 \times 1 = $ ☐

$9 \times 2 = $ ☐

$9 \times 3 = $ ☐

$9 \times 4 = $ ☐

$9 \times 5 = $ ☐

$9 \times 6 = $ ☐

$9 \times 7 = $ ☐

$9 \times 8 = $ ☐

$9 \times 9 = $ ☐

Hiroto

When the multiplier increases by 1, the answer ...

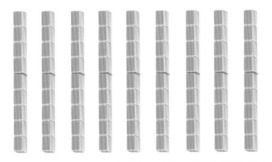

As for 9×3, decompose 9 into 5 and 4 ...

Yui

Want to try **Practice for the row of 9**

2 Let's memorize the row of 9 in the multiplication table.

2 Let's make multiplication cards for the row of 9 and practice with them.

front
9 × 3
back
27

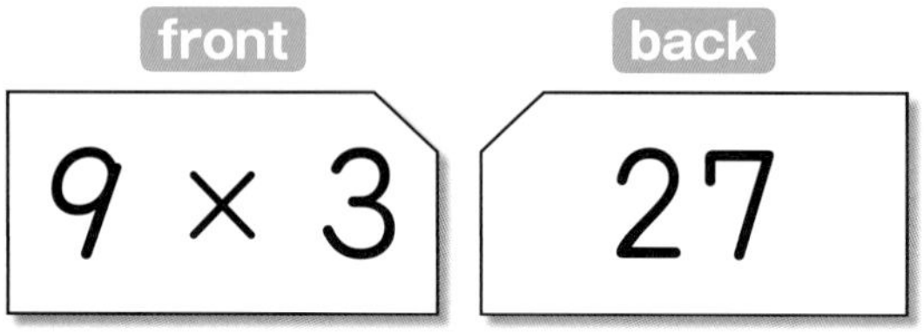

Row of 9

9 × 1 = 9	nine ones is nine
9 × 2 = 18	nine twos is eighteen
9 × 3 = 27	nine threes is twenty-seven
9 × 4 = 36	nine fours is thirty-six
9 × 5 = 45	nine fives is forty-five
9 × 6 = 54	nine sixes is fifty-four
9 × 7 = 63	nine sevens is sixty-three
9 × 8 = 72	nine eights is seventy-two
9 × 9 = 81	nine nines is eighty-one

Want to confirm

3 There are 4 buckets. Each bucket contains 9 L of water. How many liters of water are there altogether?

Want to try

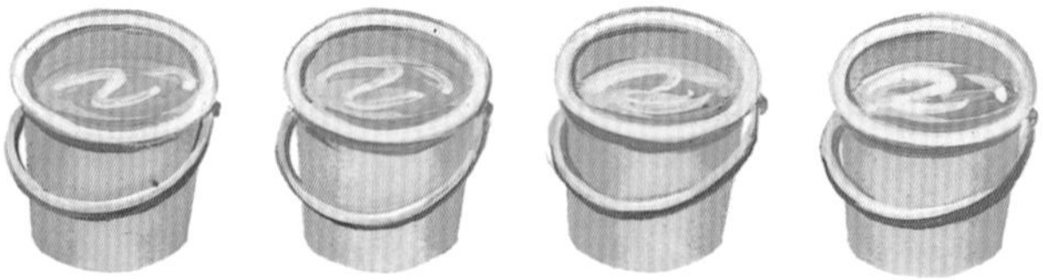

4 Let's look at the picture below and make a multiplication problem.

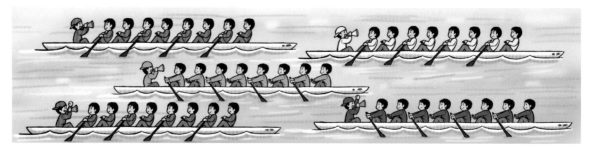

Want to solve

1 Each child receives 3 candies, two oranges, and | cake. How many of each are needed for 4 children?

① Let's think by writing a multiplication sentence.

Candies $3 \times 4 = \boxed{}$

Oranges $2 \times 4 = \boxed{}$

Cakes $\boxed{} \times \boxed{} = \boxed{}$

② Let's make the row of | in the multiplication table.

Want to try

1 Let's make multiplication cards for the row of | and practice with them.

front
back

$| \times 6$ 6

Row of |

$| \times | = |$	one ones is one
$| \times 2 = 2$	one twos is two
$| \times 3 = 3$	one threes is three
$| \times 4 = 4$	one fours is four
$| \times 5 = 5$	one fives is five
$| \times 6 = 6$	one sixes is six
$| \times 7 = 7$	one sevens is seven
$| \times 8 = 8$	one eights is eight
$| \times 9 = 9$	one nines is nine

Want to solve Various operations

1 Let's answer the following.

Way to see and think

Remember what kinds of operations you've learned so far.

① There are **8** strawberries on each dish. How many strawberries are there on **3** dishes?

② There are **9** doughnuts in each box. If **7** of them are eaten, how many doughnuts will be left?

③ There are **9** oranges in the basket and **4** oranges on the dish. How many oranges are there altogether?

④ There are **7** children and each child receives **3** pencils. How many pencils are needed?

What you can do now

☐ Can do multiplications by using the multiplication table.

1 Let's calculate the following.

① 7 × 2   ② 8 × 2   ③ 7 × 1

④ 1 × 2   ⑤ 6 × 7   ⑥ 7 × 6

⑦ 8 × 7   ⑧ 9 × 9   ⑨ 8 × 5

⑩ 1 × 5   ⑪ 7 × 3   ⑫ 6 × 8

⑬ 6 × 5   ⑭ 1 × 8   ⑮ 9 × 6

⑯ 9 × 5   ⑰ 8 × 8   ⑱ 7 × 9

⑲ 9 × 3   ⑳ 1 × 4   ㉑ 9 × 1

㉒ 7 × 7   ㉓ 6 × 3   ㉔ 8 × 1

☐ Can make a multiplication expression and find the answer.

2 There are 6 doughnuts in each box. How many doughnuts are there in 4 boxes?

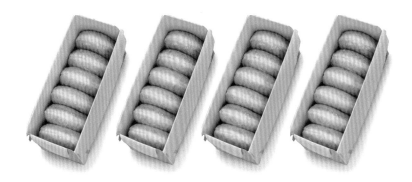

☐ Understanding what kind of operation should be used.

3 Let's represent the following ①, ②, and ③ in math sentences.

① The number of remaining chocolates when 3 chocolates were eaten from a box of 8 chocolates.

② The total number of sheets of origami paper when there are 3 bags of 8 sheets.

③ The total number of apples when there are 8 apples in the right basket and 3 in the left basket.

Supplementary Problems ▶ p.129

Usefulness and efficiency of learning

1 Let's calculate the following.

① 6×6 ② 1×3 ③ 8×4

④ 9×2 ⑤ 7×5 ⑥ 6×1

⑦ 1×7 ⑧ 6×9 ⑨ 8×9

⑩ 8×6 ⑪ 9×7 ⑫ 7×8

⑬ 1×6 ⑭ 9×8 ⑮ 1×9

Can do multiplications by using the multiplication table.

2 Let's answer the following.

① There are 8 bags of oranges. Each bag has 9 oranges. How many oranges are there altogether?

ⓐ Let's draw a diagram.

ⓑ Let's write a math expression and find the answer.

② Let's find an easier way to count the number of ●.

Can make a multiplication expression and find the answer.

ⓐ    ⓑ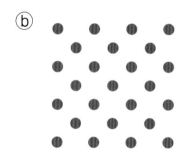

3 Let's make a math problem for 7×3.

Understanding what kind of operation should be used.

4 Let's fill in each ☐ with >, <, or =.

① 7×7 ☐ 6×8 ② 9×4 ☐ 5×9

③ 8×3 ☐ 6×4

Can do multiplications by using the multiplication table.

13 Multiplication (3)
Let's find the rules of the multiplication table and apply them.

Multiplier

	1	2	3	4	5	6	7	8	9
Row of 1 — 1									
Row of 2 — 2	4								
Row of 3 — 3									
Row of 4 — 4									
Row of 5 — 5						30			
Row of 6 — 6									
Row of 7 — 7									
Row of 8 — 8	16								
Row of 9 — 9									

Multiplicand

1 Multiplication table

Want to communicate Discovery of rules

Activity

1 Let's make the multiplication table and discover its secrets.

① Let's make the multiplication table.

② Let's make a presentation on what you have discovered.

Way to see and think

What kind of rule did you find while making the multiplication table?

Daiki's discovery

In the answer of the row of 5 in the multiplication table, the number in the ones place starts from 5, then to 0, and changes alternately.

5	10	15	20	25	30	35	40	45

Yui's discovery

There are some diagonally opposite places where the answers are the same.

25	30	35	40
30	36	42	48
35	42	49	56
40	48	56	64

Hiroto's discovery

In the row of 4, when the multiplier increases by 1, the answer increases by 4.

+1 +1 +1 +1

1	2	3	4	5	6	7	8	9
4	8	12	16	20	24	28	32	36

+4 +4 +4 +4

Nanami's discovery

The sum of the answers of the row of 2 and row of 3 equals the answer of the row of 5.

	1	2	3	4	5
1					
2	2	4	6	8	10
	+	+	+	+	+
3	3	6	9	12	15
4	↓	↓	↓	↓	↓
5	5	10	15	20	25

Want to discuss

③ Let's investigate whether the rules that the children discovered can be found in other rows.

There are some rules that we used when we made the multiplication table.

Looking at the answers in the row of 9, ...

42

2 Let's examine the rule that Yui discovered in **1**.

① Let's compare two multiplications whose answers are both 15.

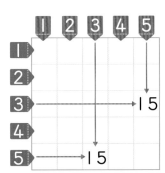

In multiplication, the answers are the same even when the order of the multiplicand and the multiplier is changed.

$$3 \times 5 = 5 \times 3$$

Let's find other pair of multiplications whose answers are the same.

② Let's fill in each ☐ with a number.

ⓐ $3 \times 8 = \boxed{} \times 3$　　ⓑ $4 \times \boxed{} = 7 \times 4$

ⓒ $\boxed{} \times 5 = 5 \times 6$　　ⓓ $9 \times 2 = 2 \times \boxed{}$

That's it! 💡

Let's try stacking one-yen coins.

The number of coins matches the answer to each multiplication. What did you notice?

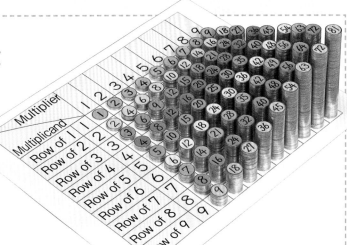

2 Using the multiplication table

Want to improve

Activity

1

How many chocolates are there altogether?

Let's find the answer in easier ways by using the multiplication table.

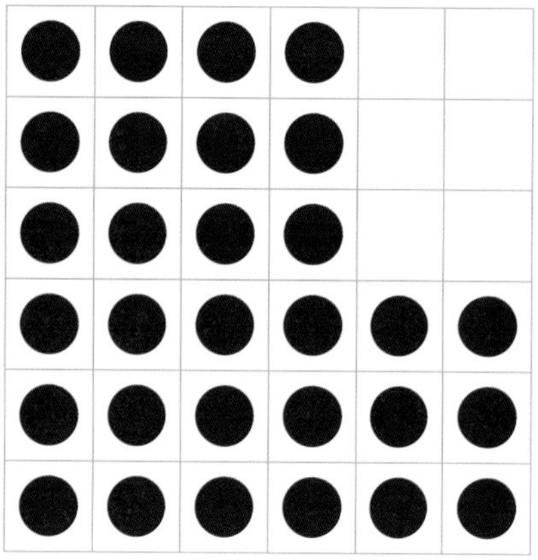

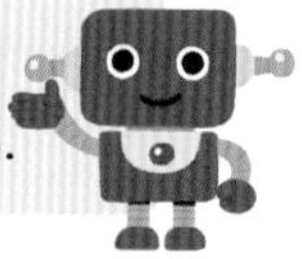

Let's represent how you found the answer by drawing lines or enclosing by circles.

Want to explain

① Let's explain the ideas of the following 4 children.

Nanami's idea

$6 \times 4 = 24$
$3 \times 2 = 6$
$24 + 6 = 30$

Answer : 30 chocolates

Hiroto's idea

$3 \times 4 = 12$
$3 \times 6 = 18$
$12 + 18 = 30$

Answer : 30 chocolates

Daiki's idea

$6 \times 6 = 36$
$3 \times 2 = 6$
$36 - 6 = 30$

Answer : 30 chocolates

Yui's idea

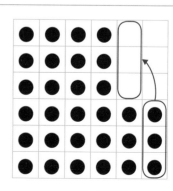

Let's represent Yui's idea in math sentences.

1 The row of 3 in the multiplication table is shown on the right.

Let's find the answer of multiplications continued from this table, such as 3×10, 3×11, and 3×12.

$3 \times 1 = 3$
$3 \times 2 = 6$
$3 \times 3 = 9$
$3 \times 4 = 12$
$3 \times 5 = 15$
$3 \times 6 = 18$
$3 \times 7 = 21$
$3 \times 8 = 24$
$3 \times 9 = 27$

Nanami's idea

In the row of 3, the answer increases by 3 from 3×1. So, the answers keep increasing even beyond 3×9.

$3 \times 9 = 27$
$3 \times 10 = 30$ $\quad) + 3$
$3 \times 11 = \boxed{}$ $\quad) + \boxed{}$
$3 \times 12 = \boxed{}$ $\quad) + \boxed{}$

Let's complete the table on page 140.

Hiroto's idea

3×5 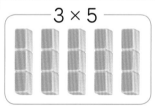 3×5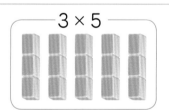

3×10

As shown in the above diagram, when the blocks are decomposed into 5 columns, the answer to 3×10 is two times the answer to 3×5.
In the same way, we can find the answer to 3×11 by decomposing 11 into 6 and 5.

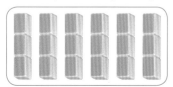 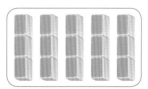

Want to connect

Thinking about how to calculate 12×3, I want to extend the multiplication table.

Hiroto

2 Hiroto and Daiki thought about how to calculate 12 × 3 as shown below.

Let's explain each idea.

 Hiroto's idea

Decompose 12 into 9 and 3.

9 × 3 = 27

3 × 3 = 9

So, 27 + 9 = 36.

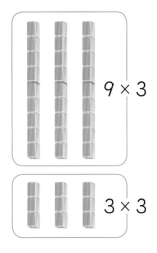

9 × 3

3 × 3

 Daiki's idea

Decompose 12 into 10 and 2.

Since 10 × 3 is three sets of 10, it is 30.

2 × 3 = 6

So, 30 + 6 = 36

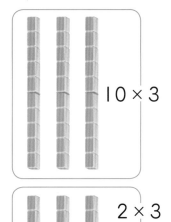

10 × 3

2 × 3

 Let's think about how to calculate 13 × 3.

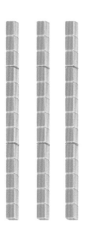

What you can do now

☐ Understanding the meaning of the multiplication table.

1 The squares shown below are parts of the multiplication table. Where would you place squares ①, ②, ③, and ④ in the parts Ⓐ, Ⓑ, Ⓒ, and Ⓓ?

　　Also, let's explain the reasons.

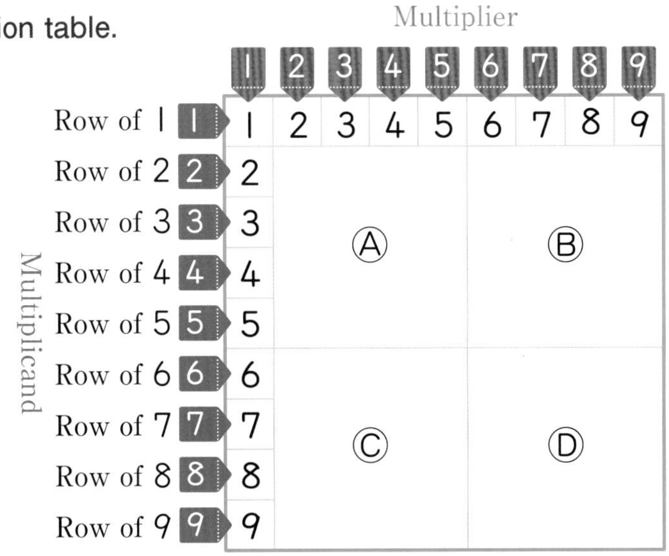

Multiplier

	1	2	3	4	5	6	7	8	9
Row of 1　1	1	2	3	4	5	6	7	8	9
Row of 2　2	2								
Row of 3　3	3		Ⓐ				Ⓑ		
Row of 4　4	4								
Row of 5　5	5								
Row of 6　6	6								
Row of 7　7	7		Ⓒ				Ⓓ		
Row of 8　8	8								
Row of 9　9	9								

Multiplicand

①

12	14	16	18
18	21	24	27
24	28	32	36
30	35	40	45

②

12	18	24	30
14	21	28	35
16	24	32	40
18	27	36	45

③

4	6	8	10
6	9	12	15
8	12	16	20
10	15	20	25

④

36	42	48	54
42	49	56	63
48	56	64	72
54	63	72	81

☐ Understanding the rules of multiplication.

2 Let's fill in each ☐ with a number.

① $3 \times 9 = 9 \times \boxed{}$　　② $7 \times 2 = \boxed{} \times 7$

③ In the row of 8, when the multiplier increases by 1, the answer increases by ☐.

☐ Can do more multiplications than 9×9.

3 The way to calculate 4×12 is shown on the right. Let's fill in each ☐ with a number.

Thinking of the row of 4,

$$4 \times 9 = 36$$
$$4 \times 10 = \boxed{}$$
$$4 \times 11 = \boxed{}$$
$$4 \times 12 = \boxed{}$$

$\Big\} + \boxed{}$
$\Big\} + \boxed{}$
$\Big\} + \boxed{}$

Usefulness and efficiency of learning

1 Let's find all the multiplication expressions that give the following answers.

① 24   ② 36

☐ Can find a math expression from its answer.

2 Let's fill in each ☐ with a number.

① In the row of 6, when the multiplier increases by 1, the answer increases by ☐.

② 9 × 3 is ☐ larger than 9 × 2.

③ 7 × 6 is ☐ smaller than 7 × 7.

☐ Understanding the rules of multiplication.

3 Yui thought about 12 × 6 as shown below.

Let's fill in each ☐ with a number.

☐ Can do more multiplications than 9 × 9.

> In multiplication, the answers are the same even when the order of the multiplicand and the multiplier is changed.
>
> 12 × 6 = 6 × 12
>
> In the row of 6, when the multiplier increases by 1, the answer increases by 6. So,
>
> 6 × 9 = ☐
> 6 × 10 = ☐
> 6 × 11 = ☐
> 6 × 12 = ☐
>
> So, the answer to 12 × 6 is ☐.

Let's deepen.

I want to know more secrets of the multiplication table.

Nanami

Deepen.

Utilize in mathematics.

Making patterns with the multiplication table

As in the row of 3, let's connect the numbers in the ones place of the answers of each row with straight lines.

Start from 0, and return to 0.

$3 \times 1 = 3$
$3 \times 2 = 6$
$3 \times 3 = 9$
$3 \times 4 = 12$
$3 \times 5 = 15$
$3 \times 6 = 18$
$3 \times 7 = 21$
$3 \times 8 = 24$
$3 \times 9 = 27$

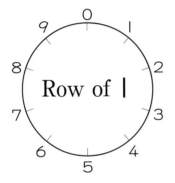

Row of 1

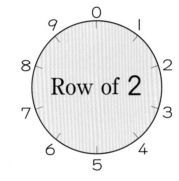

Row of 2

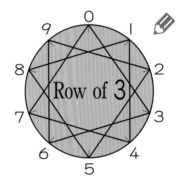

Row of 3

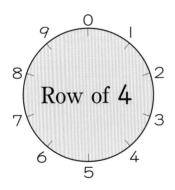

Row of 4

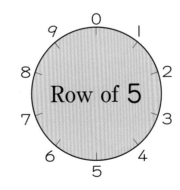

Row of 5

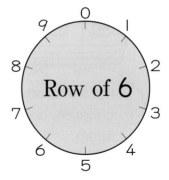

Row of 6

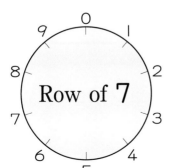

Row of 7

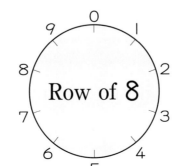

Row of 8

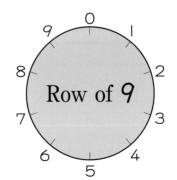

Row of 9

How to divide equally?

Problem Let's think about how to share the cake equally.

Fractions

14 Let's think by representing one part with a number.

Want to think Half of the size

1   Fold a sheet of origami paper into two parts of the same size. What ways to fold are there?

Let's think about it by drawing straight lines.

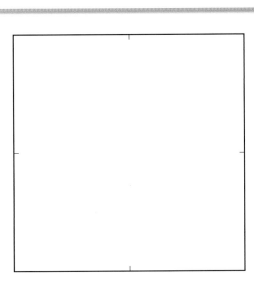
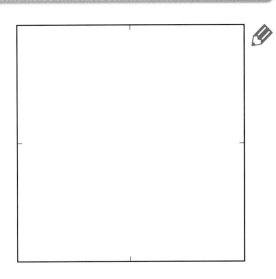

Hiroto's idea

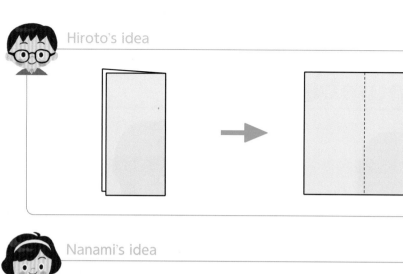

Nanami's idea

One part of an object that has been divided equally into **2** is called one half of the original and is written as $\frac{1}{2}$.

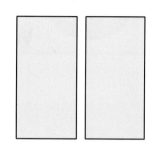

$$\frac{1}{2} \begin{array}{l} ❸ \\ ❶ \\ ❷ \end{array}$$

Want to confirm

1 ▶ Let's color $\frac{1}{2}$ of the original size.

① 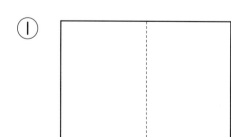   ②

2 Let's fold a sheet of origami paper two times and divide it into parts of the same size.

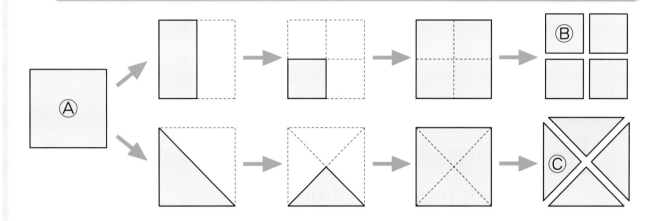

One of the parts of an object that has been divided equally into 4 is called one fourth of the original and is written as $\frac{1}{4}$.
Numbers like $\frac{1}{2}$ and $\frac{1}{4}$ are called **fractions**.

① How many pieces of Ⓑ and Ⓒ is the size of the original origami paper Ⓐ?

The size of Ⓐ is 4 times the size of Ⓑ or Ⓒ.

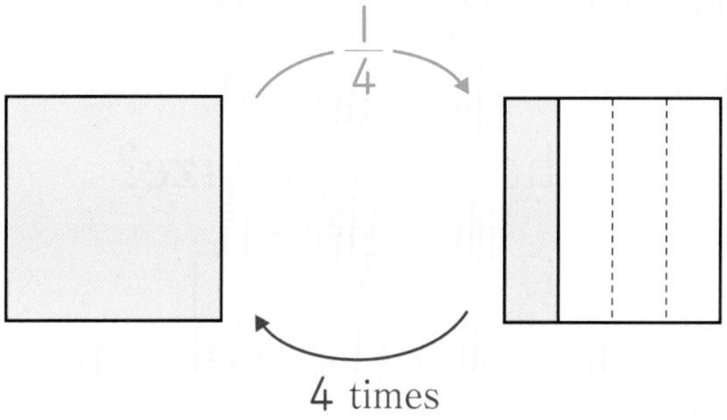

$\frac{1}{4}$

4 times

Let's color $\frac{1}{4}$ of the original size.

 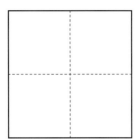

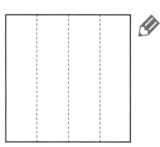

As for fractions, the new one unit is made by dividing the original size into parts of the same size.

 A sheet of origami paper was folded three times to divide it into parts of the same size.

Let's think about the following problems.

① What is the fraction of the size of Ⓐ to the size of the original origami?

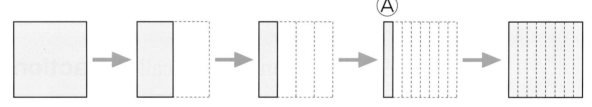

② How many times the size of Ⓐ is the size of the original origami?

What is the fraction of the size of the colored part to the original size?

 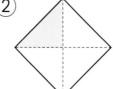

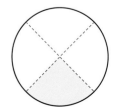

3 There is a cake on which there are 12 strawberries.
Let's cut this cake into parts of the same size.

① The cake was cut as shown below.

What is the fraction of each equally cut part of the cake to the original cake?

Hiroto's idea

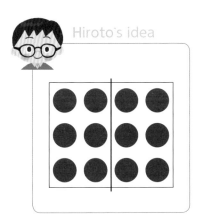

Yui's idea

Nanami's idea

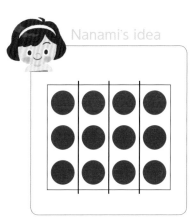

② How many strawberries are there on each equally cut part of the cake?

There are 12 strawberries on the original cake.

Daiki

Each part of cake that was cut by 3 children has a different number of strawberries.

Nanami

③ When you cut the cake in Hiroto's way, how many times the size of a part of the cake that was equally cut is the size of the original cake?

④ Also in Yui's or Nanami's way, let's think about how many times the size of a part of the cake that was equally cut the size of the original cake is.

Want to discuss

⑤ Daiki cut the cake into two pieces of the same size as shown on the right.

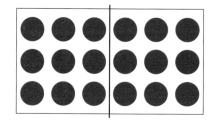

After cutting into pieces and comparing his piece to Hiroto's, he noticed the following fact.

Even though each piece is ☐ of the original size, since the sizes of the original cakes are ☐ , the size of ☐ is also different.

Let's talk about what word or number is filled in each ☐ .

What you can do now

Understanding the meaning of fractions.

1 Which of the following colored parts show $\frac{1}{2}$ of the original?

   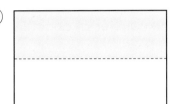   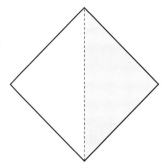

Can represent with fractions.

2 What is the fraction of the size of the colored part to the original size?

① 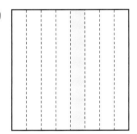   ② 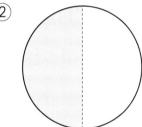   ③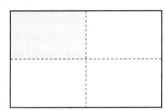

Understanding the relationship between fractions and times.

3 There are 24 balls in the box.
Then, let's think about the following problems.

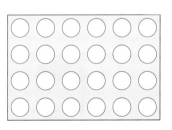

① Let's draw a straight line in the diagram on the right such that the number of balls is made $\frac{1}{3}$ of it.

8 × 3 = 24, right?

② When the number of balls is made $\frac{1}{3}$, how many balls are there?

③ How many times the number of $\frac{1}{3}$ of the balls is the original number of balls?

Yui

How can we count easily?

Want to explore How to count the number of chocolates

There are chocolates in the box. Daiki ate some of them, so the remaining chocolates are shown in the diagram on the right.

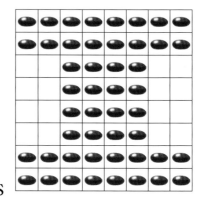

His mother said to him, "The remains will be shared with the whole family. Count how many chocolates are left." Three children heard this story and said as follows.

Yui

When the chocolates are divided by drawing a straight line, simple multiplications can be seen.
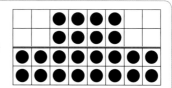

Can I move 2 chocolates projecting upward?
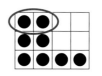

Hiroto

Nanami

If 2 chocolates can be placed into the blanks, all the grids will be filled.
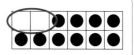

1 How many chocolates are there in the box? Also, let's write how to find the answer.

Which idea of the three children in the previous page is similar to your idea? Or, is yours an original one?

2 Let's make groups with friends that have the same idea. In the group, let's explain the way that you thought and confirm it with each other.

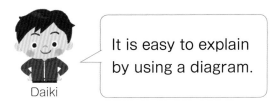

It is easy to explain by using a diagram.

Daiki

3 Let's explain the idea of your group to friends in other groups. Also, let's talk about in which way you can find the answer as easily and correctly as possible.

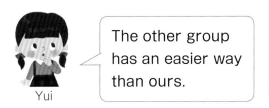

The other group has an easier way than ours.

Yui

I think the answer can be found correctly with a few number of calculations.

Hiroto

01205

59

Find the ? What time is it?

1. I'm going to play in the park.

2. I arrived earlier than usual.
   Let's play together, Yui.

3. (swinging in the park)

4. I had fun today, too.
   We played for 40 minutes.

Problem Until what time did they play in the park?

15 Time and Duration (2)
Let's tell the time and duration and find them.

Want to know Calculating time in minutes

1 Yui played with her friend for 40 minutes from 2:10 p.m.

What time did they finish playing? Let's think about it by using the following diagram.

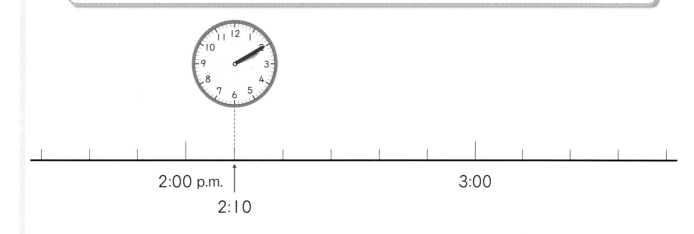

2:00 p.m.　　　　　　　　　　　　3:00

2:10

Want to confirm

 Let's find the following time.

① The time 20 minutes past 10:40 a.m.

② The time 30 minutes before 10:40 a.m.

Use a.m. or p.m. when you tell the time.

2 Let's find the following time and duration.

① The duration from **9:00** a.m. to **10:00** a.m.

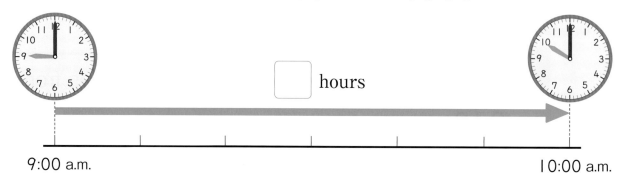

☐ hours

9:00 a.m.   10:00 a.m.

② The time **2** hours past **11:00** a.m.

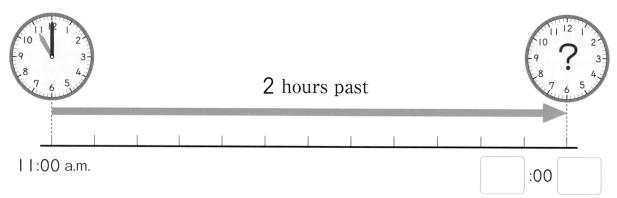

2 hours past

11:00 a.m.   ☐ :00 ☐

③ The time **1** hour before **5:00** p.m.

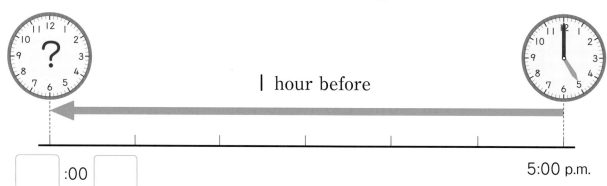

1 hour before

☐ :00 ☐   5:00 p.m.

What you can do now

☐ Can calculate time in minutes.

1 The clock on the right shows 9:15 a.m.

① What time is 20 minutes past this time?

② What time is 15 minutes before this time?

③ How many minutes does it take until 10:00 a.m.?

☐ Can calculate time in hours.

2 Let's look at the following diagrams and find the time and duration.

① The duration from 1:00 p.m. to 4:00 p.m.

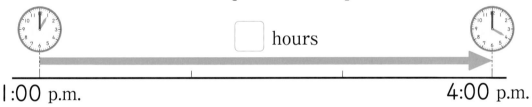

☐ hours

1:00 p.m.   4:00 p.m.

② The time 3 hours past 11:00 a.m.

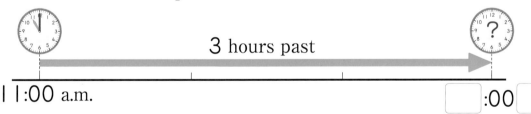

3 hours past

11:00 a.m.   ☐:00 ☐

③ The time 2 hours before 1:00 p.m.

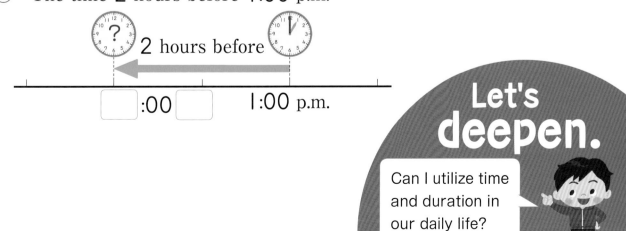

? 2 hours before

☐:00 ☐   1:00 p.m.

Let's deepen.

Can I utilize time and duration in our daily life?

Daiki

Deepen.

Field trip diary

Yuto wrote in his diary about their field trip at a zoo.

We rode a bus from school to go to the zoo for our field trip. It took us 30 minutes to go there.

At the zoo, first, our teacher talked to us at the plaza for 10 minutes, then we walked to the lion's cage for 10 minutes. After watching the lion for 20 minutes, we walked to the elephant's yard for 10 minutes.

We watched the elephant for 30 minutes and then walked back to the plaza for 10 minutes. At the plaza, I played with my friends for 1 hour.

Finally, we went back to school by bus for 30 minutes and arrived there at 12:00 noon.

What should I do to make the contents of the diary easier to understand?

Nanami

64

①   Yuto represented his field trip to the zoo by using the line of number shown below.

      Let's continue writing some more.

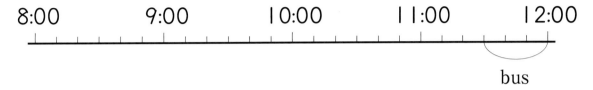

bus

②   What time did Yuto arrive at the elephant yard?

☐ : ☐

③   What time did Yuto leave school?

☐ : ☐

④   In total, how long did Yuto spend watching the animals?

☐ minutes

Hiroto

How about using the line of number that I drew in ① ?

65

Let's explore how to represent numbers and the structure.

Plastic bottle caps were collected at school.

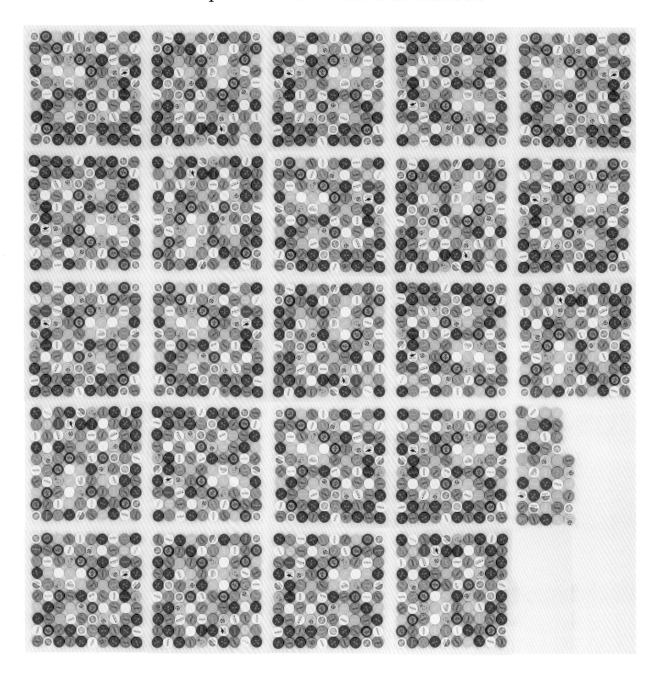

1 What is the total number of plastic bottle caps.

① Let's talk about how to count the number.

How many sets of 100 are there?

How many caps are left?

Way to see and think

10 sets of 100 make 1000.

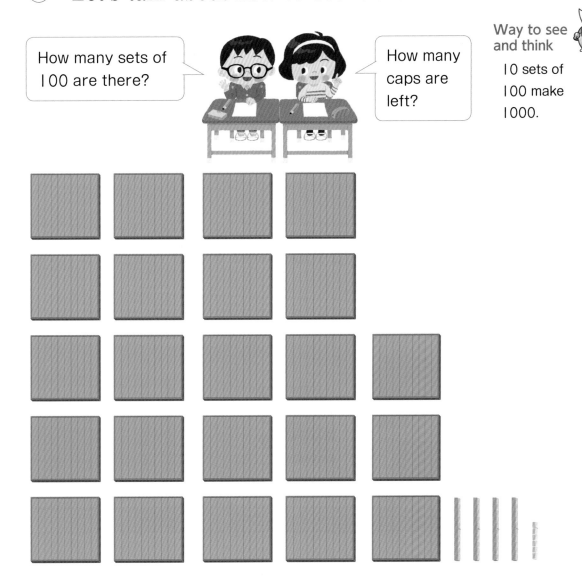

② How many sets of 1000 can be made?

⊙ Purpose How can we represent the numbers larger than 1000?

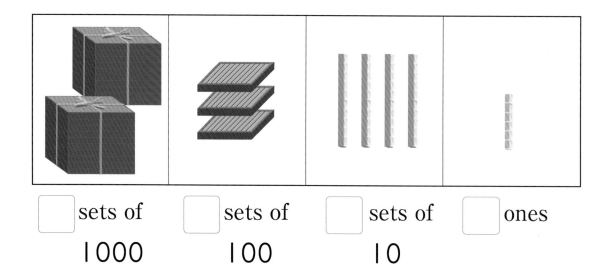

☐ sets of	☐ sets of	☐ sets of	☐ ones
1000	100	10	

🎴 Summary

The number that is the sum of two sets of 1000 is called **two thousand**.

The number that is the sum of two thousand, three hundred, forty, and six is called **two thousand three hundred forty-six** and is written as 2346.

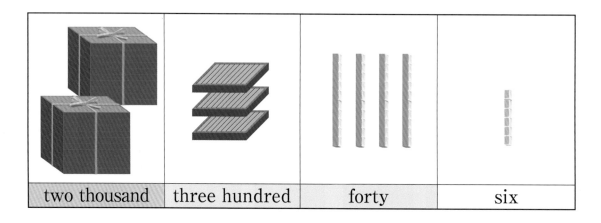

two thousand	three hundred	forty	six

Thousands place	Hundreds place	Tens place	Ones place
2	3	4	6

The position of 2 in 2346 is called the **thousands place**.

2 How many sheets of paper are there altogether?

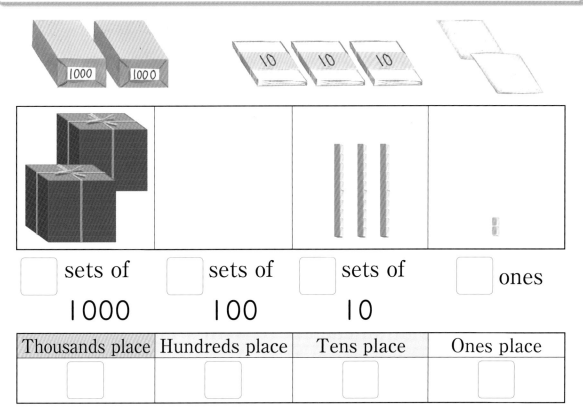

| | sets of 1000 | | sets of 100 | | sets of 10 | | ones |

Thousands place	Hundreds place	Tens place	Ones place

The number that is the sum of **2** sets of **1000**, **3** sets of **10**, and **2** ones is called two thousand thirty-two and written as **2032**.

 How many sheets of paper are there altogether?

① **3** bundles of one thousand sheets and **9** bundles of one hundred sheets.

② **5** bundles of one thousand sheets and **7** bundles of ten sheets.

Thousands place	Hundreds place	Tens place	Ones place
①			
②			

2 Let's read the following numbers.

① 6472 ② 3085

③ 1509 ④ 7003

Thousands place	Hundreds place	Tens place	Ones place	It is easier to see by using this table.

3 Let's write the following numbers.

① three thousand seven hundred forty-five

② seven thousand twenty-eight

③ three thousand one ④ five thousand

4 Let's fill in each ☐ with a number.

① The number that is the sum of 3 sets of 1000, 9 sets of 100, 2 sets of 10, and 7 ones is ☐.

② The number that is the sum of 6 sets of 1000 and 2 sets of 10 is ☐.

③ 5208 is the number that is the sum of ☐ sets of 1000, ☐ sets of 100, and ☐ ones.

④ The number that has 9 in the thousands place, 6 in the hundreds place, 4 in the tens place, and 0 in the ones place is ☐.

⑤ The number that is the sum of 4000, 200, and 7 is ☐.

3 How many sets of 100 make 2400?

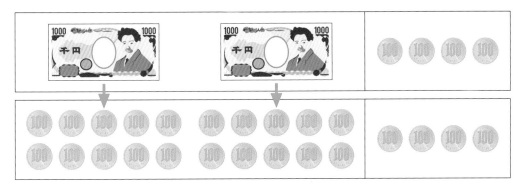

$$2400 \left< \begin{array}{l} 2000 \longrightarrow 20 \text{ sets of } 100 \\ 400 \longrightarrow 4 \text{ sets of } 100 \end{array} \right> \boxed{} \text{ sets of } 100$$

5 What is the number that is the sum of 17 sets of 100?

The number that is the sum of 10 sets of 100 is 1000.

$$17 \text{ sets of } 100 \left< \begin{array}{l} 10 \text{ sets of } 100 \longrightarrow 1000 \\ 7 \text{ sets of } 100 \longrightarrow 700 \end{array} \right> \boxed{}$$

6 How many sets of 1000 make 8000?

Also, how many sets of 100 make 8000?

4 How many • are there in the diagram below?

① Let's circle each set of 1000 dots.

② How many sets of 1000 are there?

The number that is the sum of 10 sets of 1000 is called **ten thousand** and is written as 10000.

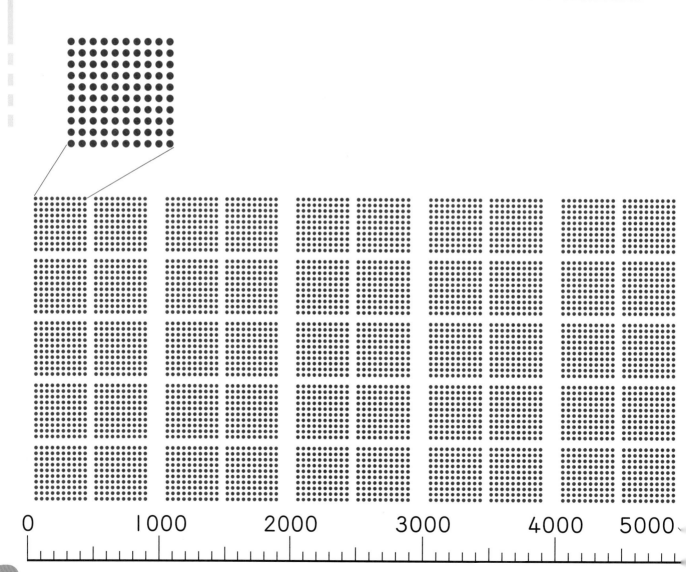

0   1000   2000   3000   4000   5000

7 Let's answer by using the line of number below.

Way to see and think

Let's think about what the smallest scale on the line represents.

① What is the number that is added to 9000 to get 10000?

② What is the number that is 100 smaller than 10000?

③ What is the number that is 1 smaller than 10000?

④ How many sets of 100 make 10000?

⑤ Let's draw ↑ at the scale on the line that represents 4500.

⑥ Let's draw ↑ at the scale on the line that represents 9800.

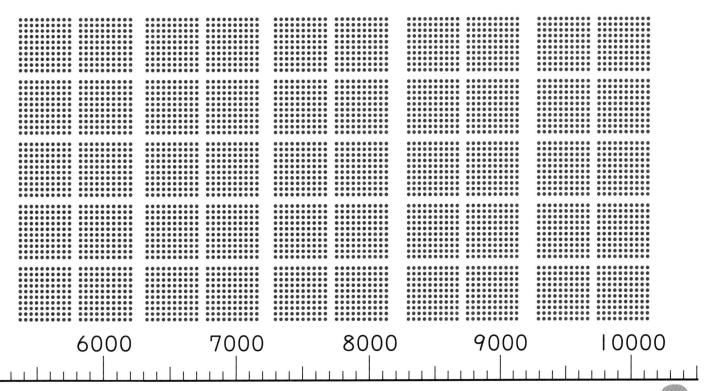

6000　　　7000　　　8000　　　9000　　　10000

5 Let's fill in each ☐ with a number.

① — 5000 – 6000 – ☐ – 8000 – ☐ – ☐ –

② — 7500 – ☐ – ☐ – 9000 – 9500 – ☐ –

③ — ☐ – 9996 – 9997 – ☐ – 9999 – ☐ –

④ — 6400 – ☐ – 6800 – 7000 – ☐ – ☐ –

8 Let's draw lines to connect the following numbers in ascending order.

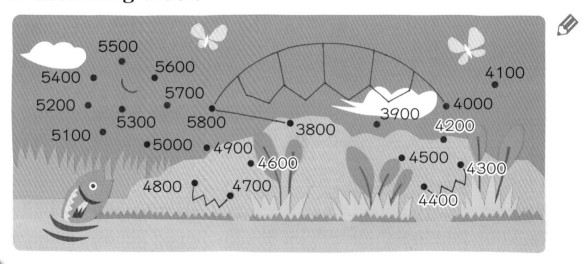

9 Let's answer by using the line of number below.

① Let's read the numbers that are pointed by ↑ at ⓐ, ⓑ, and ⓒ.

② Let's draw ↑ at the scale on the line that represents **3200**.

0   1000   2000   3000   4000   5000

        ↑   ↑   ↑

      ⓐ   ⓑ   ⓒ

10 Let's write the following numbers.

① The number that is **800** larger than **3200**.

② The number that is **300** smaller than **3200**.

③ The number that is **80** larger than **5820**.

④ The number that is **1000** smaller than **10000**.

⑤ The number that is **100** larger than **9900**.

Want to represent Comparing the size of numbers

Way to see
and think

In which place
should we
compare the
numbers?

 11 Which number is larger?

Let's fill in the ☐ with > or <.

① 4950 ☐ 5190

4900 5000 5100 5200

4950   5190

Thousands	Hundreds	Tens	Ones

② 8340 ☐ 8610

8300 8400 8500 8600

Thousands	Hundreds	Tens	Ones

③ 9253 ☐ 9238

9220 9230 9240 9250

Thousands	Hundreds	Tens	Ones

What you can do now

Understanding how to represent numbers.

1 How many sheets of paper are there altogether?

Understanding the structure of numbers.

2 For the number **7620**, let's fill in each ☐ with a number.

① The number that is the sum of ☐ sets of **1000**,

☐ sets of **100**, and ☐ sets of **10**.

② The number that is the sum of **7000, 600,** and ☐ .

③ The number that is the sum of ☐ sets of **10**.

④ The number that is ☐ larger than **7600**.

Understanding the size of 4-digit numbers.

3 Let's write the following numbers.

① The number that is **1000** larger than **5000**.

② The number that is **200** smaller than **7000**.

③ The number that is **460** larger than **2000**.

Understanding the size of 4-digit numbers.

4 Let's draw ↑ at the scale on the line that represents numbers ①, ②, ③, and ④.

① **5800**　② **6300**　③ **8900**　④ **9900**

5000　　6000　　7000　　8000　　9000　　10000

Supplementary Problems
p.133

Usefulness and efficiency of learning

1 Let's write the following numbers.

Understanding how to represent numbers.

① The number that is the sum of 8 sets of 1000, 4 sets of 100, and 6 ones.

② The number that is the sum of 3 sets of 1000 and 6 sets of 10.

2 For the number 5080, let's fill in each ☐ with a number.

Understanding the structure of numbers.

① 5 in the thousands place means that there are 5 sets of ☐ .

② The number that is the sum of 5000 and ☐ .

3 Two cards protrude from the respective envelopes as shown below. In what situation can you say that number Ⓑ is larger than number Ⓐ?

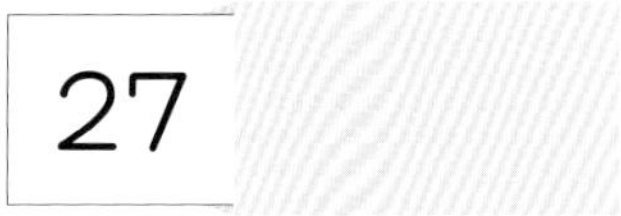

Understanding the size of 4-digit numbers.

Ⓐ 98   Ⓑ 27

That's it! 💡 Numbers in our surroundings

Numbers are used in various places in our surroundings.
Let's find where numbers are used.

Room number of a hotel

←1401~1411 **14F** 1412~1422⇒

1422 means room 22 on the 14th floor.

It's a number for identifying a room.

Nanami   Hiroto

The number on the license plate of a motorcycle is also an identification number.

License plate of a motorcycle

今治市 しまなみ A 2011

Reflect Connect

Problem

Let's color grids on the grid paper and make beautiful patterns.

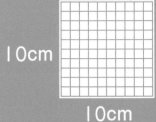
10cm

10cm

How many grids are there in one sheet? ⟶ 100 grids

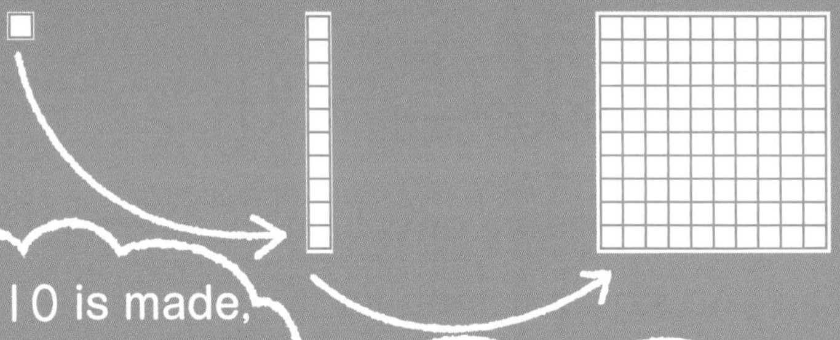

1   10   100

If a set of 10 is made, carry 1 to the next higher place.

10000 grids make a square with a length of 1 m and a width of 1 m.

There are 10000 grids in 100 sheets of paper that have 100 grids each.

If there are 96 sheets of paper that have 100 grids each, $10 \times 9 + 6 = 96$

In total, 9600 grids ⟵

Let's decide how to color by ourselves. I'll color every other grid.

Daiki

I want to make patterns such as those reflected in a mirror.

Yui

Let's arrange the colored squares. | How many grids are there in total?

1000 grids are enclosed.

Close! It's 400 less than 10000.

9 sets of 1000 make 9000.
(10 sets make 10000.)

6 sets of 100 make 600.

Hiroto

How many grids are the sum of the grids that every child colored?

There are 96 sheets in total.

Nanami

Want to connect

Daiki

Are there numbers larger than 10000? How should I read those numbers?

Find the ? What is the length of outstretched arms?

Problem Let's think about how to measure longer lengths.

17 Length (2)

Let's think about how to compare longer lengths and represent them.

Longer lengths

1 Soyoka measured the length of her outstretched arms by using a tape. She found out that it was 3 times with a 30-cm ruler and 25 cm more.

Let's answer the following.

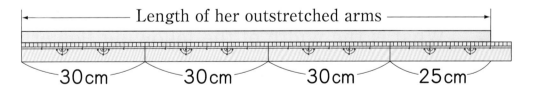

Length of her outstretched arms

30cm   30cm   30cm   25cm

① What is the length of Soyoka's outstretched arms in cm?

It's longer than 100 cm.

It is convenient to use a longer ruler.

Yui

Daiki

81

100 cm is called 1 **meter** and is written as 1 m.

m is also a unit for length.

$$1m = 100cm$$

❶ ❷ ❸
1 m

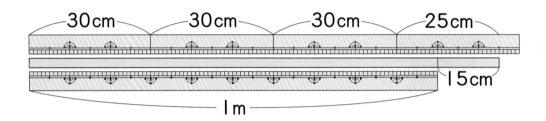

It is convenient to use a 1-m ruler when you measure longer lengths.

② What is the length of Soyoka's outstretched arms in m and cm?

115cm = ☐ m ☐ cm

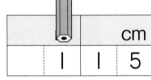

		cm
1	1	5

m		cm
1	1	5

Want to confirm

What is the width of the following flowerbed in m and cm?

Also, how many cm is it?

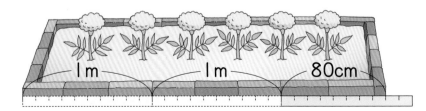

2 Let's cut a tape by guessing the length of 1 m.

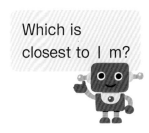

Which is closest to 1 m?

 Let's measure the length of things in our surroundings.

(1) First, let's estimate the length.

(2) Then, let's measure the actual length.

3 There are a 3 m 20 cm tape and a 2 m tape.

Let's answer the following.

— 3m 20cm —   — 2m —

① Let's find the sum of the lengths of the two tapes.

$$3 \text{ m } 20 \text{ cm} + 2 \text{ m}$$

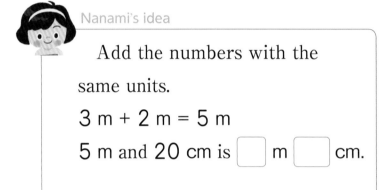

Nanami's idea

Add the numbers with the same units.

3 m + 2 m = 5 m

5 m and 20 cm is ☐ m ☐ cm.

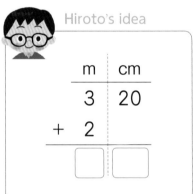

Hiroto's idea

m	cm
3	20
+ 2	
☐	☐

② Let's find the difference between the lengths of the two tapes.

3 There are a 12 m string and a 9 m string.

What is the sum of the lengths of the two strings in m? Also, what is their difference in m?

4 Let's calculate the following.

① 8 m + 15 m   ② 15 m 50 cm − 7 m

What you can do now

☐ Understanding how to represent longer lengths.

1 Let's fill in each ☐ with a number.

① 1 m is the length that is the sum of 10 sets of ☐ cm.

② The length that is 3 sets of 1 m is ☐ m.

③ The sum of 2 m and 30 cm is ☐ m ☐ cm.

Also, it is ☐ cm.

☐ Can compare lengths.

2 Let's fill in each ☐ with >, <, or =.

① 12 cm ☐ 12 m

② 6 m 7 cm ☐ 607 cm

☐ Can calculate length.

3 Let's answer the following.

① Let's calculate the following.

ⓐ 5 m 60 cm + 3 m   ⓑ 2 m 40 cm + 15 m

ⓒ 8 m 20 cm − 4 m   ⓓ 7 m 52 cm − 30 cm

② There are two tapes.

Ⓐ 20 cm   Ⓑ   1 m 50 cm

ⓐ What is the sum of the lengths of two tapes?

ⓑ Which tape is longer, Ⓐ or Ⓑ, and by how much?

ⓒ Let's find the length when 2 tapes Ⓐ and 1 tape Ⓑ

are added.

Supplementary Problems p.135

Usefulness and efficiency of learning

1 Let's find the lengths of tapes.

Understanding how to represent longer lengths.

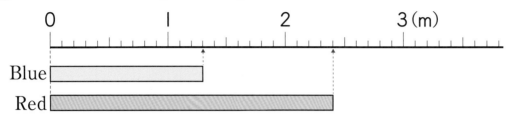

① What are the lengths of the blue and red tapes in m and cm?

② What are the lengths of the blue and red tapes in cm?

2 Let's arrange the following lengths from the longest.

Can compare lengths.

① 3 m　　　　310 cm　　　2 m 80 cm

② 4 m 50 cm　　405 cm　　　4 m 15 cm

3 Ren measured the width and height of a bookshelf.

Can calculate length.

· Width : once with a 1-m ruler and 20 cm more

· Height: once with a 1-m ruler, twice with a 30-cm ruler, and 10 cm more

① What are the width and height of the bookshelf in m and cm?

② Which is longer, the width or the height, and by how many cm?

Let's deepen.

I want to make a tape of which length represents a unit of length to measure different lengths.

Hiroto

Deepen.

Utilize in life.

Let's measure with tapes!

Want to know

A tape was cut into tapes Ⓐ, Ⓑ, and Ⓒ as shown below.

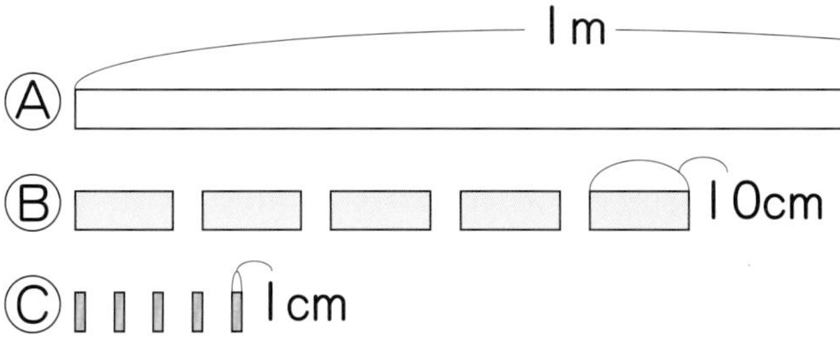

Ⓐ ⎯⎯ 1 m ⎯⎯   1 piece

Ⓑ ▭ ▭ ▭ ▭ ▭ 10cm   5 pieces

Ⓒ ‖‖‖‖‖ 1cm   5 pieces

Want to represent

① When the width of a table was measured by using these tapes, the result became as shown in the diagram below.

What is the width of the table in m and cm?

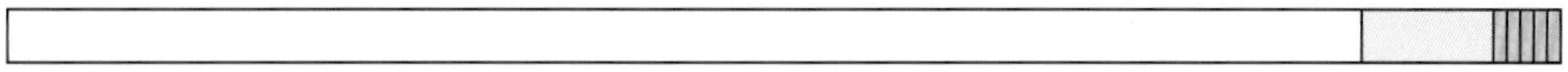

Want to explain

② When the length of the table was measured by using 1 piece of Ⓐ, 1 piece of Ⓑ, and 2 pieces of Ⓒ, it was 88 cm.

Let's explain by using words and math expressions how it was measured.

Problem

Let's explore the lengths in our surroundings.

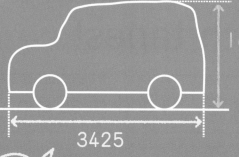

1675

3425

The length of the car is represented in mm.

How much is a length of 3425 mm?

Let's represent the relationship between mm, cm, and m in the diagram and organize it.

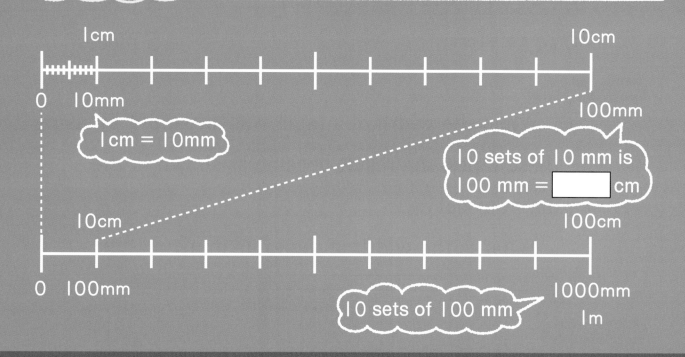

1cm   10cm

0   10mm   100mm

1cm = 10mm

10 sets of 10 mm is
100 mm = ☐ cm

10cm   100cm

0   100mm   1000mm
                                               1m

10 sets of 100 mm

How much is
a length of
3425 mm?

Daiki

1cm = 10mm
1m = 100cm
Right?

Hiroto

Let's represent the
relationship between
mm, cm, and m in
the diagram and
organize it.

88

> Since 10 sets of 100 mm is 1000 mm,
>   1000 mm = ☐ cm = ☐ m

Let's think by decomposing 3425 mm.

3000 mm = ☐ cm = ☐ m

400 mm = ☐ cm

20 mm = ☐ cm

5 mm

In total, 3425 mm = ☐ m ☐ cm 5 mm

Summary

10 mm = 1 cm, 100 cm = 1 m

So, 1000 mm = 1 m

m		cm	mm
3	4	2	5

Let's deepen.

① The height of this car is 1675 mm. What is it in m, cm, and mm?
② Let's compare the size of the blackboard in your classroom and this car.

Yui: The height of this car is taller than our height.

Nanami: Is it larger than the blackboard?

Let's measure the width and height of the blackboard and represent them in mm.

How many people came on board?

There are 27 passengers on the bus.

Several more people came on board later.

There are 34 passengers in total.

How many people came on board?

Problem Let's think about how to find the number of people that came on board later.

Addition and Subtraction

18 Let's think about how to calculate by using diagrams.

1 There are 12 red marbles and 14 blue marbles. There are 26 marbles altogether.

Let's represent this situation by using a diagram.

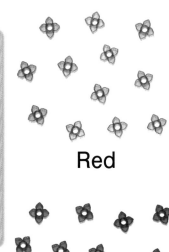

Red

Blue

Nanami's diagram

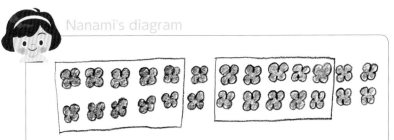

Hiroto's diagram

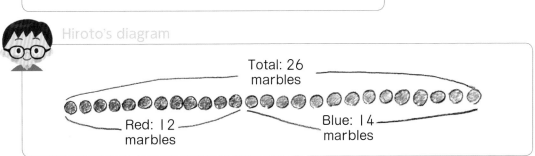

Total: 26 marbles

Red: 12 marbles

Blue: 14 marbles

Yui's diagram

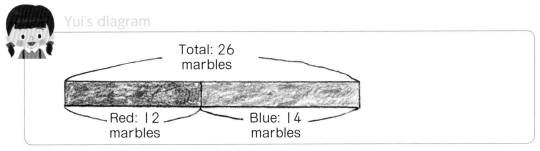

Total: 26 marbles

Red: 12 marbles

Blue: 14 marbles

① Let's talk about the good points of each diagram.

There are **38** sheets of blue paper and **63** sheets of red paper.

How many sheets of paper are there altogether?

Blue   Red

① Let's represent this problem in order from (1) to (3) with diagrams.

(1) **38** sheets of blue paper

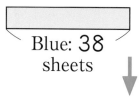

Blue: **38**
sheets

It's all right even if the partitions are not exact.

(2) **63** sheets of red paper

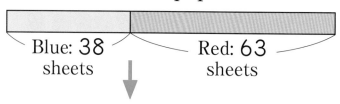

Blue: **38**
sheets

Red: **63**
sheets

(3) How many sheets of paper are there altogether?

Write the number that you don't know as □.

Total: ☐ sheets

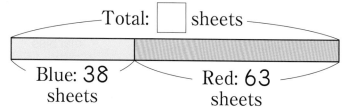

Blue: **38**
sheets

Red: **63**
sheets

② Let's write a math expression and the answer.

 There are **21** red pencils and **23** blue pencils.

How many pencils are there altogether?

2 In the 2nd term, there were 29 children in Hitomi's class. 3 new children joined the class in the 3rd term.

How many children are there altogether?

① Let's represent this problem in order from (1) to (3) with diagrams.

(1) 29 children in the 2nd term.

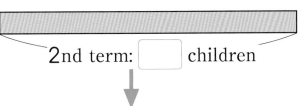

2nd term: ☐ children

Way to see and think

Let's think about what should be represented as ☐.

(2) 3 children joined.

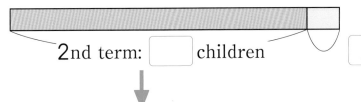

2nd term: ☐ children   ☐ children joined

(3) How many children are there altogether?

Total: ☐ children

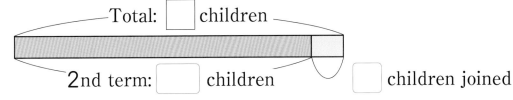

2nd term: ☐ children   ☐ children joined

② Let's write a math expression and the answer.

3 There were 18 oranges. I received 12 more oranges. How many oranges are there altogether?

3 There was a 23 m tape. You used 17 m of it.

How many m of the tape are left?

① Let's represent this problem in order from (1) to (3) with diagrams.

Way to see and think

Let's think about what should be represented as □.

(1) There was a 23 m tape.

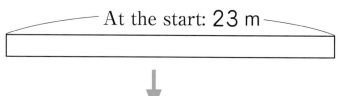

At the start: 23 m

(2) 17 m of it were used.

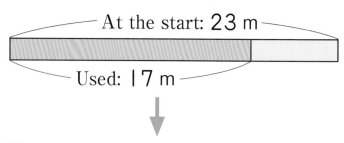

At the start: 23 m

Used: 17 m

(3) How many m are left?

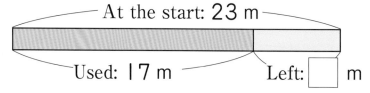

At the start: 23 m

Used: 17 m   Left: ☐ m

② Let's write a math expression and the answer.

Want to try

4 There are 50 sheets of colored paper. You used 16 sheets of them. How many sheets are left?

4 Class 1 has 29 children and Class 2 has 31 children. What is the difference of the number of children?

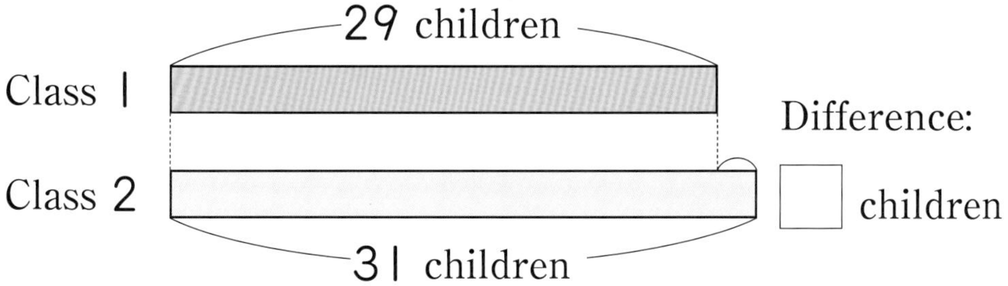

It's easier to compare by drawing a diagram with two tapes.

29 children

Class 1

Class 2

Difference:

☐ children

31 children

① How many fewer children are there in Class 1 than in Class 2?

② How many more children are there in Class 2 than in Class 1?

5 Nanami folded 18 cranes. Yui folded 23 cranes. What is the difference of the number of cranes?

① Let's complete the following diagram.

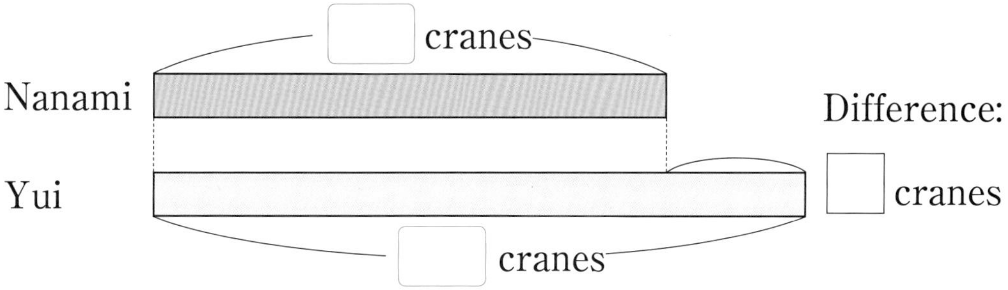

☐ cranes

Nanami

Yui

Difference:

☐ cranes

☐ cranes

② Who folded more cranes and by how many?

5
I picked up 18 acorns. My teacher picked up 4 more acorns than me.
How many acorns did the teacher pick up?

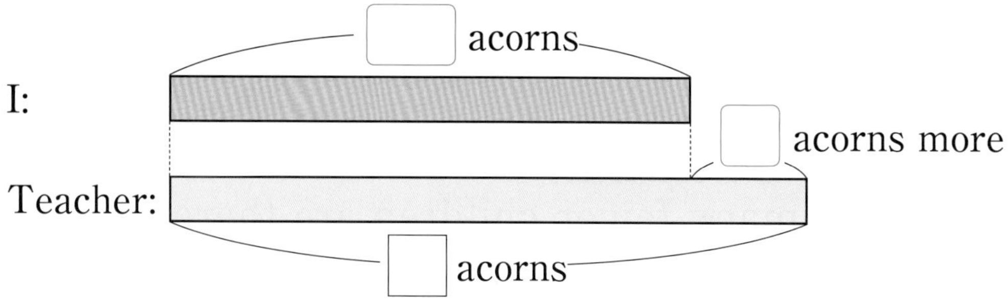

I: [   ] acorns

[   ] acorns more

Teacher: [   ] acorns

6   There are 32 red tulips. There are 8 more white tulips than red tulips.
How many white tulips are there?

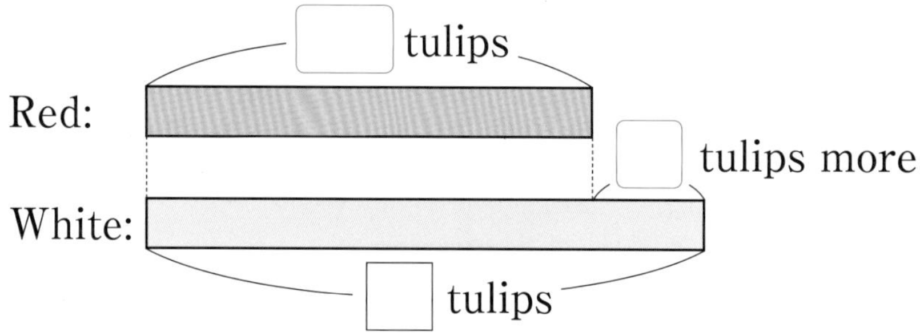

Red: [   ] tulips

[   ] tulips more

White: [   ] tulips

6

I picked up 3 1 empty cans. Sota said he picked up 5 fewer cans than me.

How many cans did Sota pick up?

```
                    ┌──┐ cans
  ┌──────────┐  ┌────────────────────────┐
  │          │  │░░░░░░░░░░░░░░░░░░░░░░░░░│
  └──────────┘              ┌──┐ cans fewer
  ┌──────────┐  ┌──────────────────┐
  │          │  │                  │
  └──────────┘     ┌──┐ cans
```

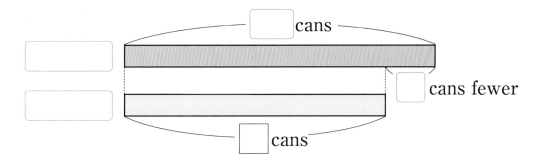

7

There are 23 girls in the classroom. There are 7 fewer boys than girls.

How many boys are there?

```
              ┌──┐ children
Girls: ┌────────────────────────┐
       │░░░░░░░░░░░░░░░░░░░░░░░░░│
                  ┌──┐ children fewer
Boys:  ┌──────────────────┐
       │                  │
          ┌──┐ children
```

7 We took a group photo. There were 8 chairs, where one child sat on each, and 13 children stood.

How many children were in the photo?

8 chairs

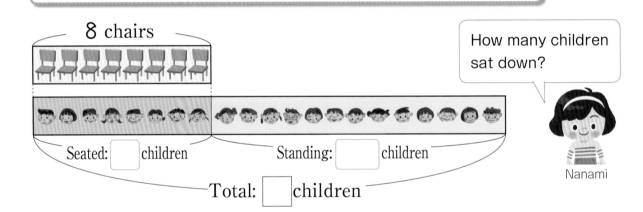

Seated: ☐ children   Standing: ☐ children

Total: ☐ children

How many children sat down?

Nanami

8 There are 15 cakes. When there are 6 plates with one cake each, how many cakes are left?

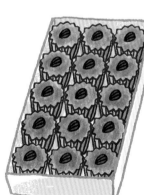

6 plates

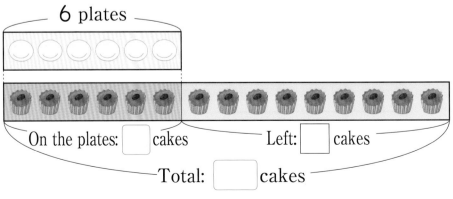

On the plates: ☐ cakes   Left: ☐ cakes

Total: ☐ cakes

8 There were 27 passengers on the bus. Several more people came on board later, and then there were 34 passengers in total.

How many people came on board?

① Let's draw diagrams that match the situations below.

(1) **27** passengers on the bus.

At the start:
27 people

27

(2) Several people came on board.

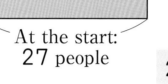

At the start: Later: ☐ people
27 people

27 + ☐

(3) The total number of passengers became **34**.

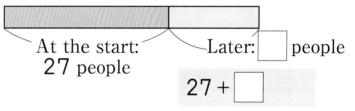

Total: 34 people

At the start: Later: ☐ people
27 people

27 + ☐ = 34

② Let's write a math expression to find the answer, and then write the answer.

9 ▶ Mei had some marbles. She gave **6** marbles to her sister. She counted the remaining number, and **18** marbles were left.

How many marbles did she have at the start?

① Let's draw diagrams that match the situations below.

(1) Mei had some marbles.

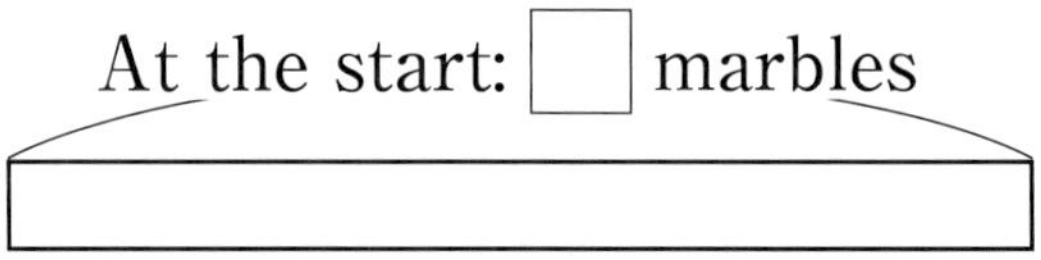

At the start: ☐ marbles

☐

(2) She gave **6** marbles to her sister.

At the start: ☐ marbles

Gave:
6 marbles

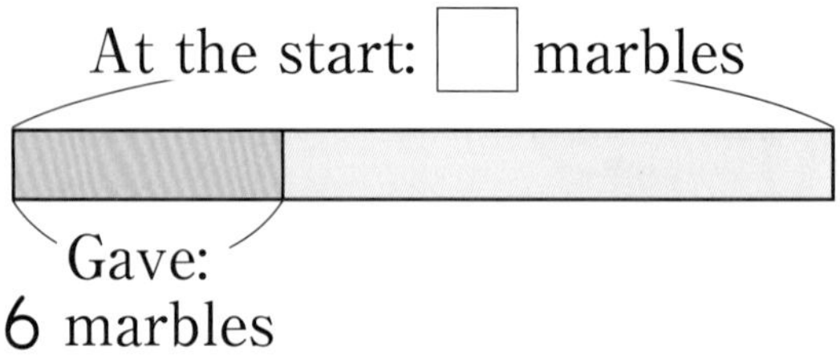

☐ − 6

(3) She counted the remaining number, and **18** marbles were left.

At the start: ☐ marbles

Gave: Left:
6 marbles   18 marbles

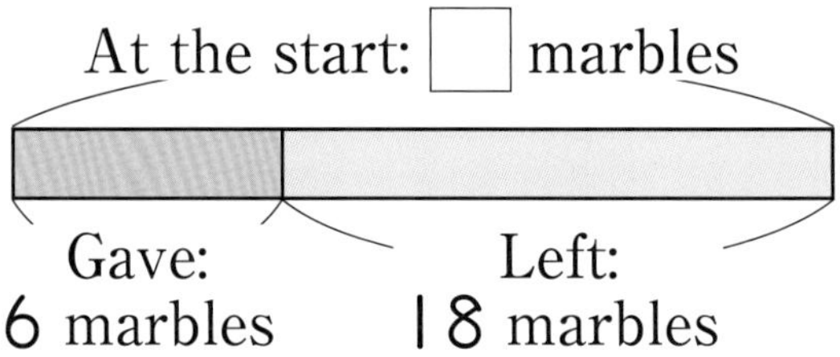

☐ − 6 = 18

② Let's write a math expression to find the answer, and then write the answer.

9 Tsubasa had 110 stickers. Since he gave some of them to his friend, 83 stickers were left.

How many stickers did he give to his friend?

① Let's draw diagrams that match the situations below.

(1) Tsubasa had 110 stickers.

At the start: 110 stickers

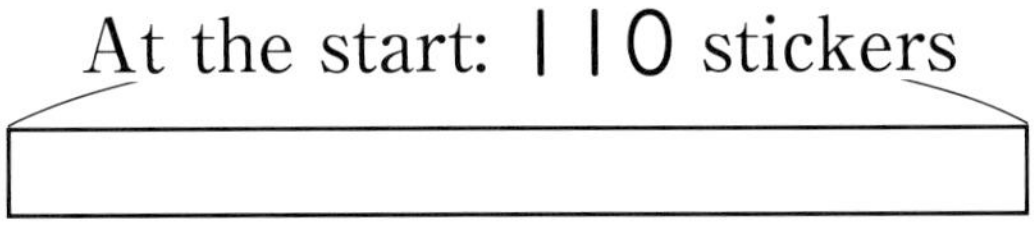

110

(2) He gave some of them.

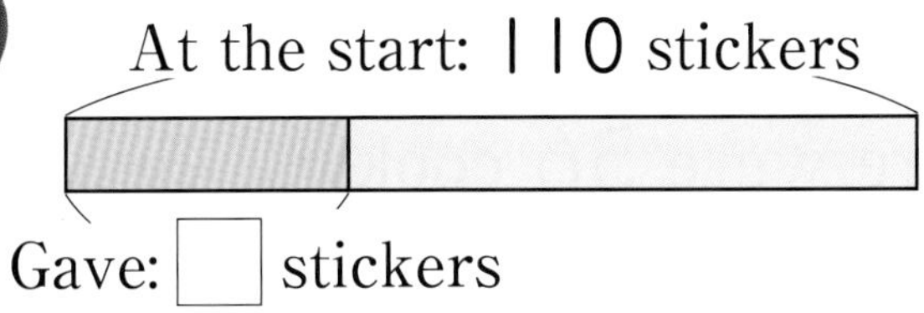

At the start: 110 stickers

Gave: ☐ stickers

110 − ☐

(3) 83 stickers were left.

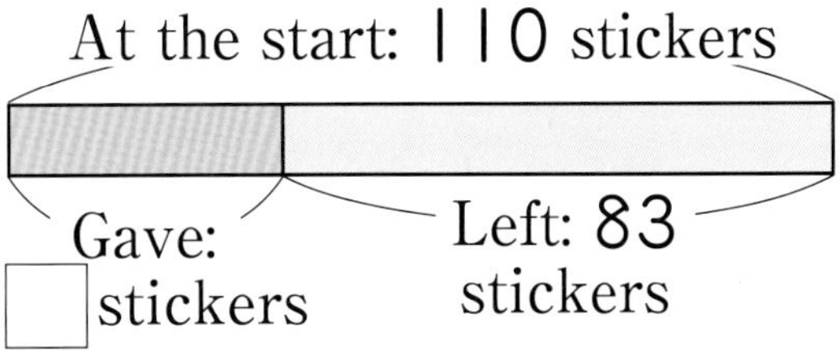

At the start: 110 stickers

Gave: ☐ stickers   Left: 83 stickers

110 − ☐ = 83

② Let's write a math expression to find the answer, and then write the answer.

There were children playing in the ground. Since **26** children joined later, there are **42** children in total.

How many children were there at the start?

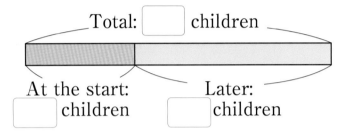

Total: ☐ children

At the start:
☐ children

Later:
☐ children

① Let's write a math expression to find the total number.

② Let's write a math expression to find the answer, and then write the answer.

There were **56** cookies. Since some of them were eaten, **29** cookies were left.

How many cookies were eaten?

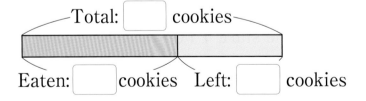

Total: ☐ cookies

Eaten: ☐ cookies   Left: ☐ cookies

① Let's write a math expression to find the number of remaining cookies.

② Let's write a math expression to find the answer, and then write the answer.

Want to think Making a math problem

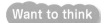

10 There are 17 blue ribbons and 24 red ribbons.

There are 41 ribbons altogether.

Let's make a math problem by using these numbers.

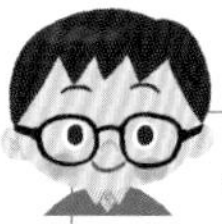

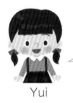

Hiroto's problem

There are 17 blue ribbons and 24 red ribbons.

How many ribbons are there altogether?

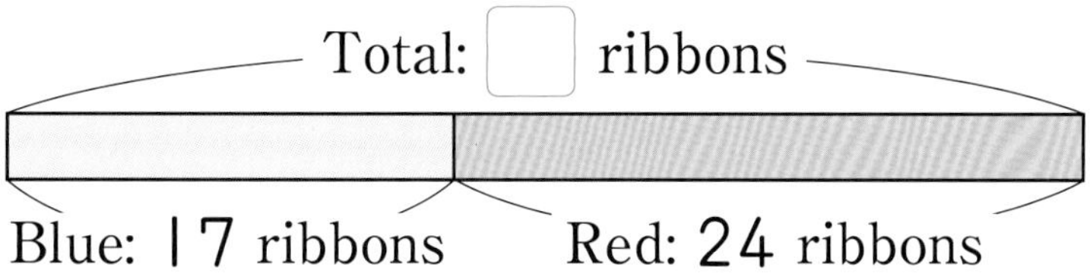

Total: ☐ ribbons
Blue: 17 ribbons
Red: 24 ribbons

Yui

I'll make a math problem to find the number of blue ribbons.

Way to see and think

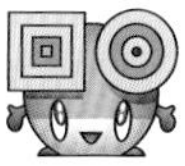

By changing numbers and words in the diagram, you can make various addition and subtraction problems.

Want to confirm

12 Let's look at the following diagrams and make math problems.

①

Total: 15 flowers
Gave: ☐ flowers
Left: 9 flowers

②

Total: 18 pieces
At the start: 8 pieces
Bought: ☐ pieces

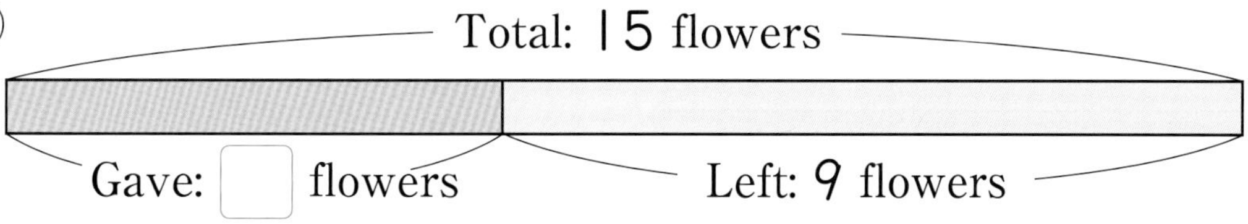

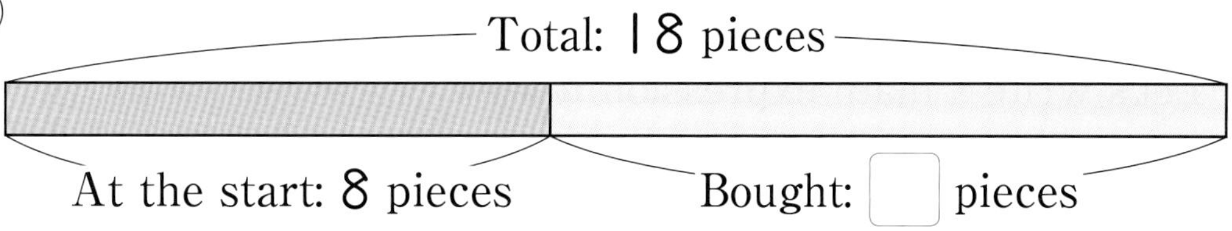

What you can do now

☐ Understanding the relationship between the situation and diagram.

1 A diagram that matches the following situation is shown below.

Let's fill in each ( ) with a phrase from ☐ and complete the diagram. Also, let's find the answer.

> There are 31 killifish in the left aquarium and 18 killifish in the right aquarium.
>
> How many killifish are there altogether?

> Right aquarium: 18 killifish   Total: ☐ killifish
>
> Left aquarium: 31 killifish

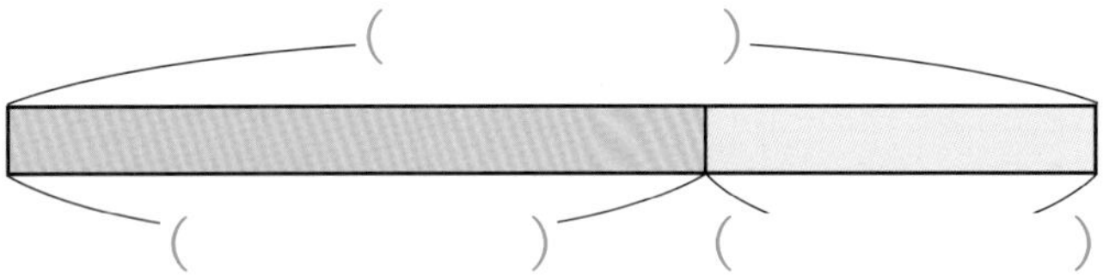

```
        (         )
   ┌─────────────┬──────────┐
   │█████████████│          │
   └─────────────┴──────────┘
     (        )   (            )
```

☐ Can make a math expression and find the answer.

2 Hayato had several cards. Since his brother gave him 13 cards, he has 42 cards now. How many cards did Hayato have at the start?

① Let's look at the following diagram and write a math expression to find the total number of cards.

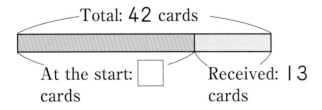

② Let's write a math expression to find the answer, and then write the answer.

Supplementary Problems p.135

Usefulness and efficiency of learning

1 Let's look at the following diagrams and fill in each ☐ with a number and in each ( ) with a word.

① Saki has ☐ marbles. Yamato has ☐ ( ) marbles than her. How many marbles does ( ) have?

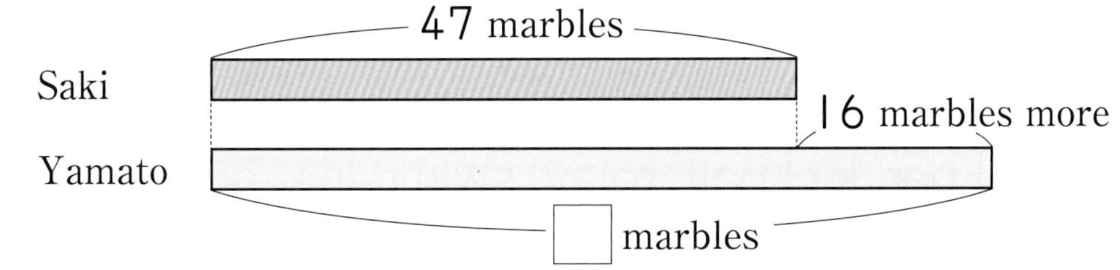

```
                  ┌─── 47 marbles ───┐
Saki    ┌─────────────────────────────┐
        └─────────────────────────────┘
                                    ┐ 16 marbles more
Yamato  ┌────────────────────────────────┐
        └────────────────────────────────┘
              └──── ☐ marbles ────┘
```

② There were ☐ plastic bottles of tea. ☐ of them were handed out. How many bottles are ( )?

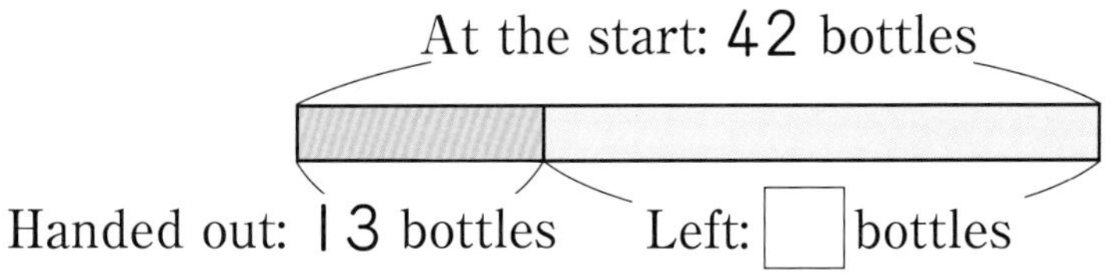

```
              ┌── At the start: 42 bottles ──┐
        ┌────────┬──────────────────┐
        └────────┴──────────────────┘
Handed out: 13 bottles   Left: ☐ bottles
```

2 53 children borrowed books from a library in October. The number of children is 24 fewer than that in September. How many children borrowed books in September?

Can make a math expression and find the answer.

① Let's look at the following diagram and write a math expression to find the number of children who borrowed books in October.

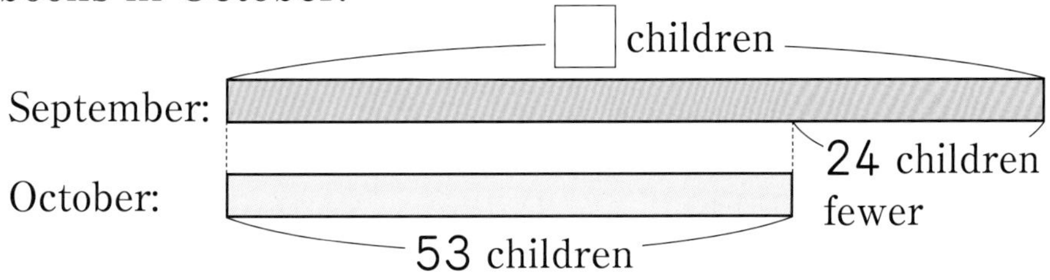

```
                    ┌── ☐ children ──┐
September: ┌──────────────────────────────┐
          └──────────────────────────────┘
                                  ┐ 24 children
October:  ┌────────────────────┐    fewer
          └────────────────────┘
          └──── 53 children ────┘
```

② Let's write a math expression to find the answer, and then write the answer.

19 Let's think about how to arrange and organize data.

1 All children of Hiroto's class were asked what sport they want to do in their P.E. class from the following: horizontal bar, kickball, relay, swimming, and dancing. The result is shown below.

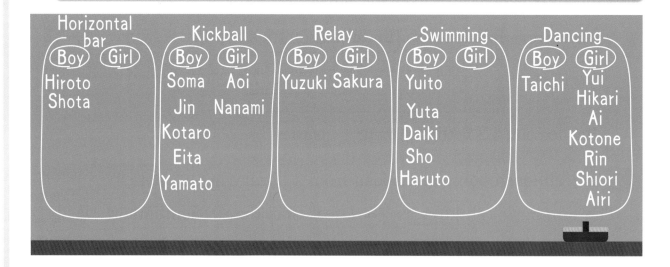

① Let's write the number of children for each sport in the following table.

What children want to do in P.E. class

Sport	Horizontal bar	Kickball	Relay	Swimming	Dancing
Number of children					

Want to think

② Which sport has the largest number of children that want to do it? How many children want to do it?

Want to represent

③ Let's represent the number of children for each sport by using ○ on the following graph in descending order.

What children want to do in P.E. class

That's it!

What do you see when you summarize two things into one graph?

The table "What children want to do in P.E. class" summarized in **1** is divided into those for boys and girls. Let's summarize the two tables into the following graph and talk about what you noticed.

What children want to do in P.E. class (Boys)

Sport	Horizontal bar	Kickball	Relay	Swimming	Dancing
Number of children	2	5	1	5	1

What children want to do in P.E. class (Girls)

Sport	Horizontal bar	Kickball	Relay	Swimming	Dancing
Number of children	0	2	1	0	7

What children want to do in P.E. class

Boys	Girls	Boys	Girls	Boys	Girls	Boys	Girls	Boys	Girls
Horizontal bar		Kickball		Relay		Swimming		Dancing	

What is the shape of a box?

Problem What are the properties of the shape of boxes?

20 Shapes of Boxes
Let's explore the shape of the flat parts of boxes.

Shapes of boxes

1 Let's explore the shapes of boxes Ⓐ and Ⓑ.

Ⓐ

Ⓑ

① Let's trace the flat part of the box.

Let's mark so as to see what part is traced.

 The flat part of a box is called a **face**.

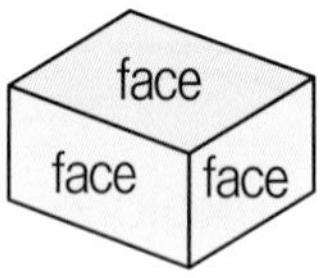

Though there are various shapes of boxes, are the numbers of faces the same?

Nanami

Daiki

What is the shape of the faces?

Ⓨ **Purpose** What is the shape of the faces of a box and how many faces are there?

② Let's cut out each face.

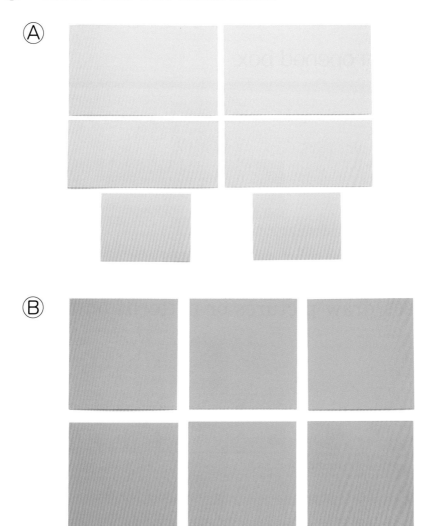

Ⓐ

Ⓑ

③ What kind of quadrilateral is the shape of each face?

④ How many faces are there?

Are the numbers or shapes of the faces different according to boxes?

Summary

The shapes of the faces of the boxes are rectangles and squares, and the number of faces is 6.

2 Let's connect the faces by using tape and create the figure of an opened box.

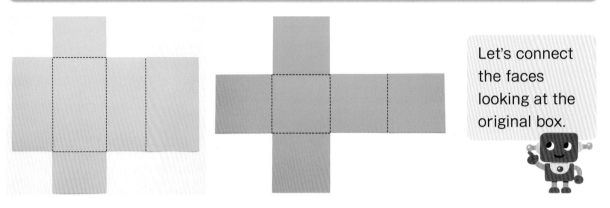

Let's connect the faces looking at the original box.

① Let's draw pictures or patterns on the figure.

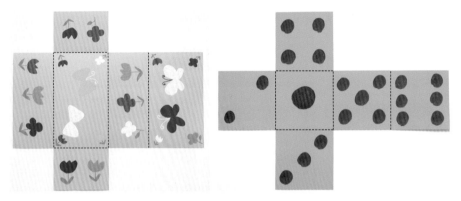

② Let's fold the paper of the figure to make a box.

 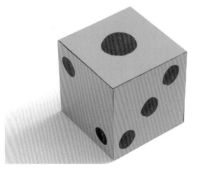

When a box is folded, what are the opposite faces?

③ Let's look at the box and talk about what you noticed.

1 Let's circle the set of shapes which becomes a box when it is formed. As for the set which doesn't become a box, let's say the reason.

Way to see and think

Way to see and think

What are the properties of the shape of boxes?

Ⓐ (   )

Ⓑ (   )

Ⓒ (   )

Ⓓ (   )

3 Let's make a box as shown on the right with sticks and balls of clay. Let's explore the shape of this box.

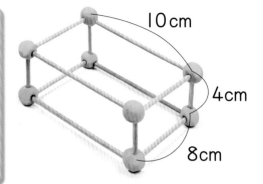

10 cm

4 cm

8 cm

① How many cm of sticks and how many sticks are needed?

10 cm: ☐ sticks   8 cm: ☐ sticks   4 cm: ☐ sticks

② How many balls of clay are needed?

Each stick on the shape of the box is called an **edge**.

Each ball of clay is called a **vertex**.

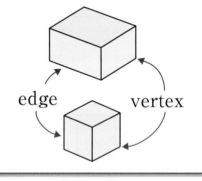

edge   vertex

③ How many edges and vertices are there in the shape of the box?

2 Let's make a box as shown on the right with sticks and balls of clay. How many sticks are needed? How many balls of clay are needed?

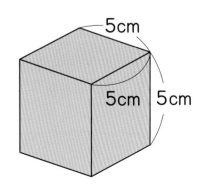

5 cm

5 cm   5 cm

What you can do now

☐ Understanding the number of faces, edges, and vertices of the shape of a box.

1 Let's make a shape of the box as shown on the right with sticks as edges and balls of clay as vertices.

Let's answer the following.

① How many balls of clay are needed?

② How many 12 cm sticks are needed?

③ How many 6 cm sticks are needed?

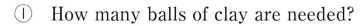

☐ Understanding what faces are needed to make a box.

2 When you fold the following figures ①, ②, and ③, which box is formed respectively? Let's connect with lines.

① 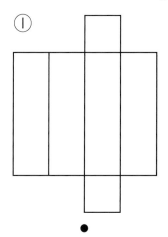   ② 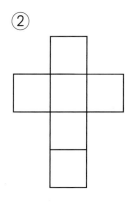   ③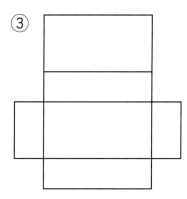

Ⓐ 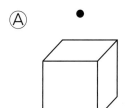   Ⓑ 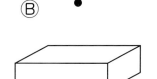   Ⓒ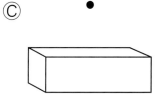

Usefulness and efficiency of learning

1 As for the shapes of the following boxes, how many faces, edges, and vertices are there respectively?

Understanding the number of faces, edges, and vertices of the shape of a box.

Ⓐ

Ⓑ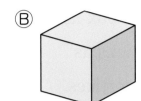

Faces: ☐

Edges: ☐

Vertices: ☐

Faces: ☐

Edges: ☐

Vertices: ☐

2 As for the figure below, one face is missing to make a box by folding.

Let's add the face that is needed to make the box.

Understanding what faces are needed to make a box.

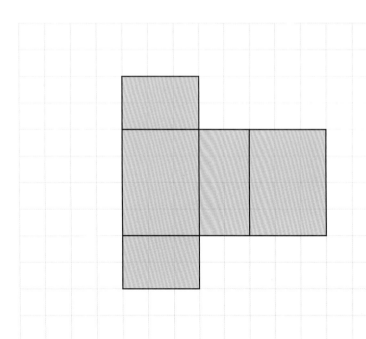

Let's deepen.

Can I make a box-shaped thing?

Yui

Deepen.

Utilize in life.

Let's make a tool box.

Want to know

Let's make a tool box which fits into the shelf shown on the right.

Let's draw all the faces that are needed to make the box below.

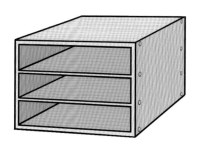

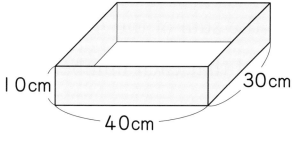

10cm

30cm

40cm

A box without a top will be made.

10cm

10cm

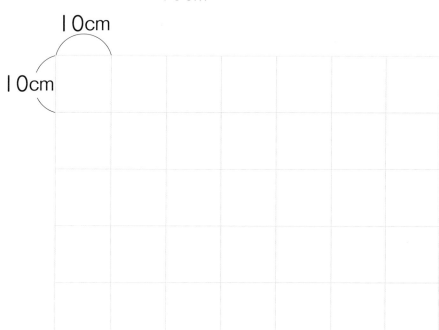

Large numbers

1 Let's make the largest number and the smallest number by using the 4 cards: 1, 3, 5, and 9.

Yui

Addition and subtraction

2 Let's fill in the ☐ with a number from 1, 2, 3, 4, 5, 6, and 7 such that the sum of the 4 numbers in each circle is the same and that each number appears only one time.

Example   Sum 13

14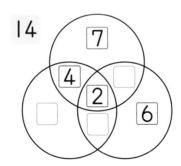

15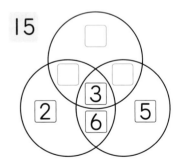

16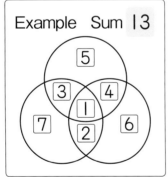

17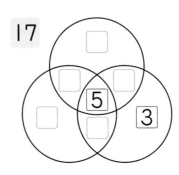

18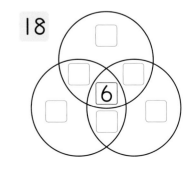

3 Let's fill in each ☐ with a number.

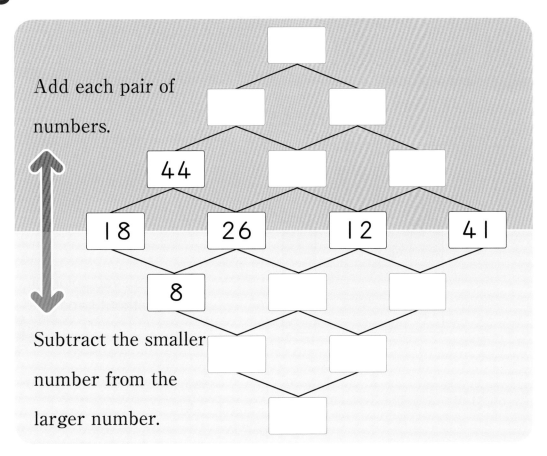

Add each pair of numbers.

Subtract the smaller number from the larger number.

Multiplication

4 Let's write **4** multiplication sentences where all the answers have different numerals.

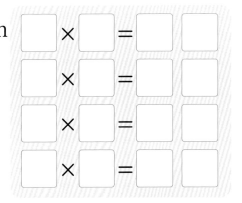

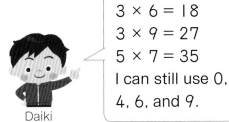

3 × 6 = 18
3 × 9 = 27
5 × 7 = 35
I can still use 0, 4, 6, and 9.

Daiki

8 × 8 = 64, right?

Nanami

Trace some of the dotted lines to make quadrilaterals. The numbers correspond to the number of grids.

Looking at the example, let's make quadrilaterals and color them.

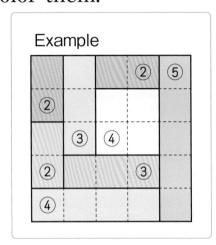

Example

The space that is marked with ④ is a shape of 1×4, 2×2, or 4×1.

Can you color all without gaps?

①

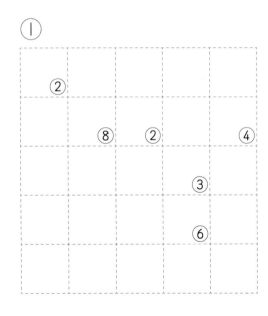

②

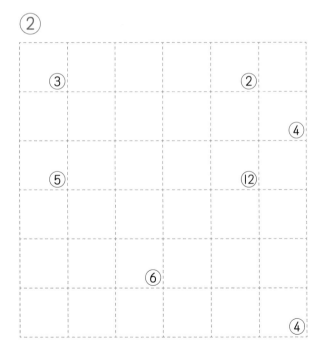

Let's make original questions. After that, let's change questions with a friend and try to solve it.

Triangle and quadrilateral

6 Let's draw three different triangles and three different quadrilaterals by connecting dots with straight lines.

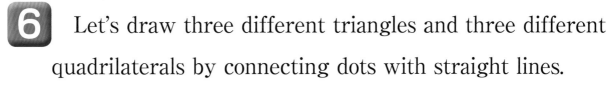

Length

7 How long are the following straight lines in cm and mm? Also, how long are they in mm?

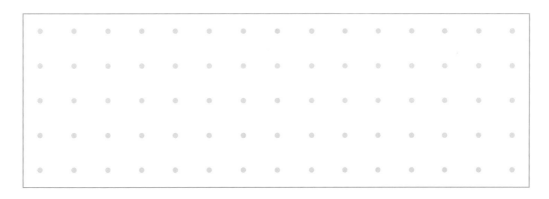

Amount of water

8 The amount of water in each container is measured. How many L and dL of water are there? Also, how many dL of water are there?

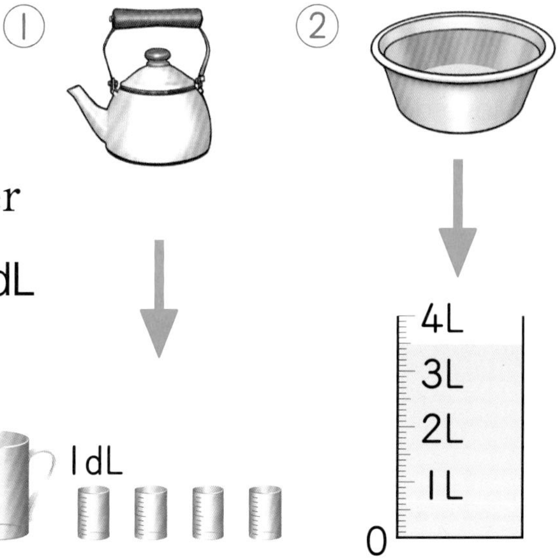

Computational thinking

01206

Let's teach Robo "a method to move the disks of the Tower of Hanoi."

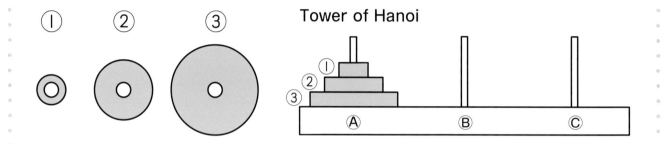

Tower of Hanoi

Rules:

- Move disks ①, ②, and ③ on rod Ⓐ onto rod Ⓑ.
- You can move only one disk at a time, but you cannot place any larger disk on top of a smaller disk.

① When you give the following instructions to Robo, he will move disks respectively. Let's try to explain.

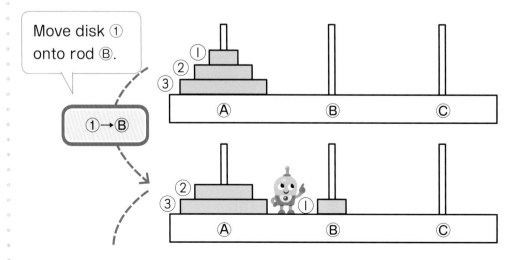

Move disk ① onto rod Ⓑ.

① → Ⓑ

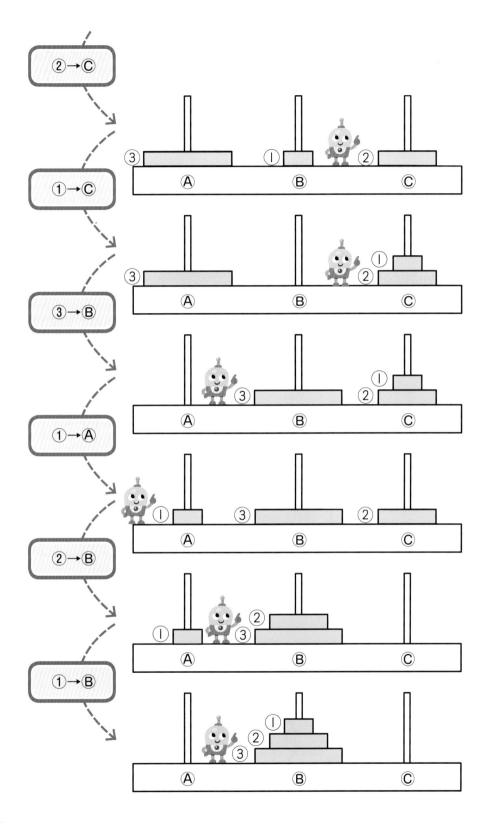

②   Let's teach Robo how to move the disks on rod Ⓑ
onto rod Ⓒ.

Let's make a quiz about our school or area.

I want new second grade students to know much more things about our school and town.

I remember we received the present "school and town quiz" when we became second grade students.

I want to make an enjoyable quiz.

Let's make quizzes about what we've learned in the second grade.

They were divided into 4 teams according to content they had learned, and then each team made a quiz.

1. Numbers and calculations team

On the bulletin board, you can put up 6 sheets of drawing paper in height and 8 sheets in width. How many sheets can you put up on the board altogether?

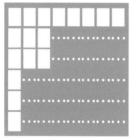

? Let's try making a quiz in which multiplication is used.

2. Measurement team

Clock Ⓐ shows the time when you leave home to go to school. Clock Ⓑ shows the time when you come home. How many hours does it take from the time you leave home to the time you come home?

Ⓐ

Ⓑ

I want to make a quiz about a beetle I caught during the summer vacation and the number of its legs.

Daiki

I want to make a quiz about the number of all second grade students.

Nanami

I want to make a quiz about the length of the horizontal bar.

Hiroto

I want to make a quiz about places where we can walk in 20 minutes from our school.

Yui

? Let's try making a quiz in which measuring is used.

3. Figures team

In our school, let's look for shapes that are like the shapes of ▢ , ▭ , and ◺ .
Let's take pictures of them and create an album with the pictures.

The shape of the door of the classroom is a rectangle.

We can find different kinds of shapes outside school.

Nanami

? Let's try making a quiz about a shape we can see in our surroundings.

4. Graph team

Yui

I want to investigate in which month the largest number of children were born and make a quiz about it.

I want to investigate what sport they like in my class and make a quiz about it.

Daiki

The graph on the right shows the number of children in a class of the second grade who grew each kind of vegetables.

Let's guess what vegetable is in ▢ .

Let's investigate in your class what vegetable your classmates want to grow.

Vegetables we grew in our class

	Cucumbers	Bitter gourds	Peppers	Soybeans
●				
●	●			
●	●			
●	●	●	●	
●	●	●	●	●
●	●	●	●	●

? Let's investigate what you are interested in and represent it in a graph. Also, let's make a quiz about it.

What I can do now.

1. Toward learning competency

	😊 Strongly agree	😐 Agree	☹ Don't agree
① It was fun making a quiz.			
② The learning contents were helpful.			
③ I made a quiz on my own initiative.			

2. Thinking, deciding and representing competency

	😊 Definitely did	😐 So so	☹ I didn't
① I was able to discover a quiz in which mathematics is used.			
② I was able to confirm whether I can solve the quiz.			
③ I was able to represent quizzes with words, pictures, and figures.			

3. What I know and can do

	😊 Definitely did	😐 So so	☹ I didn't
① I was able to make a better quiz.			
② I was able to solve the quiz I made.			

4. Encouragement for myself

	😊 Strongly agree
① I think that I'm doing my best.	

Give yourself a compliment since you have worked so hard.

Let's try to work out what you were not able to accomplish and keep doing your best on what you were able to fulfill.

Supplementary Problems

⑪ Multiplication (1)

pp.4~28

❶ Let's represent the following blocks in multiplication expressions and find the total number.

①

②

③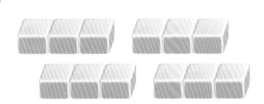

❷ Let's fill in the ☐ with a number.

The answer to 6 × ☐ is the same as the answer to

6 + 6 + 6 + 6.

❸ How many times the length of tape ▭ are the lengths of the following tapes?

① ▭

② ▭

❹ There are 3 boxes stacked together and each box is 4 cm high.

What is the total height in cm?

❺ Let's calculate the following.

① 5 × 3   ② 5 × 1

③ 5 × 2   ④ 5 × 6

⑤ 5 × 5   ⑥ 5 × 7

⑦ 5 × 9   ⑧ 5 × 8

❻ There is a square with a side of 5 cm. What is the length around the square in cm?

7 Let's calculate the following.
① 2×3 ② 2×5
③ 2×2 ④ 2×1
⑤ 2×4 ⑥ 2×9
⑦ 2×8 ⑧ 2×7

8 There are 6 children and each child is given 2 sheets of origami paper. How many sheets are needed altogether?

9 Let's calculate the following.
① 3×2 ② 3×4
③ 3×8 ④ 3×1
⑤ 3×3 ⑥ 3×7
⑦ 3×6 ⑧ 3×9

10 3 children can sit on one couch. How many children can sit on 5 couches?

11 Let's calculate the following.
① 4×1 ② 4×3
③ 4×2 ④ 4×6
⑤ 4×8 ⑥ 4×5
⑦ 4×9 ⑧ 4×7

12 There are 4 dumplings in each stick. How many dumplings are there in 4 sticks?

12 Multiplication (2)
pp.29~40

1 Let's calculate the following.
① 6×4 ② 6×3
③ 6×1 ④ 6×6
⑤ 6×2 ⑥ 6×9
⑦ 6×5 ⑧ 6×7

2 There are 6 oranges in each bag.
   How many oranges are there in 8 bags?

3 Let's calculate the following.
① 7×3 ② 7×5
③ 7×2 ④ 7×1
⑤ 7×9 ⑥ 7×7
⑦ 7×4 ⑧ 7×8

4 There are 6 boxes that contain 7 pieces of gum each.

How many pieces of gum are there altogether?

5 Let's fill in each ☐ with a number.

① In the row of 7, when the multiplier increases by 1, the answer increases by ☐.

② The answer to 8×6 is ☐ larger than the answer to 8×5.

6 Let's calculate the following.

① 8×2 ② 8×5
③ 8×6 ④ 8×1
⑤ 8×7 ⑥ 8×9
⑦ 8×4 ⑧ 8×8

7 There are 3 packs that contain 8 eggs each.

How many eggs are there altogether?

8 Let's calculate the following.

① 9×4 ② 9×2
③ 9×5 ④ 9×3
⑤ 9×9 ⑥ 9×7
⑦ 9×6 ⑧ 9×1

9 Each child was given a 9 cm ribbon. How many cm of ribbon are needed for 8 children?

10 Let's fill in each ☐ with a number.

The answer to 9×2 plus the answer to 9×3 equals ☐.

This is the same as the answer to $9 \times$ ☐.

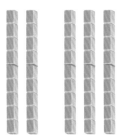

11 Let's calculate the following.

① 1×5 ② 1×2
③ 1×3 ④ 1×6
⑤ 1×8 ⑥ 1×4
⑦ 1×7 ⑧ 1×9

⑬ Multiplication (3)

pp.41～49

1 Let's make the multiplication table. Let's answer the following.

Multiplier

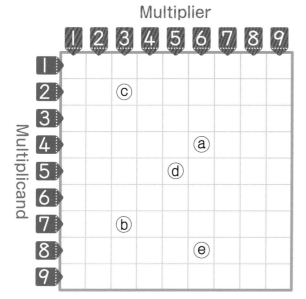

Multiplicand

① The answers to what multiplication expressions are ⓐ and ⓑ?

② What are the appropriate numbers for ⓒ, ⓓ, and ⓔ?

③ In the row of 3 and 5, by how many does each answer increase when the multiplier increases by 1?

2 Let's connect the math expressions with the same answer.

| 7 × 8 | 6 × 7 | 8 × 9 | 9 × 6 |

| 6 × 9 | 9 × 8 | 7 × 6 | 8 × 7 |

3 Let's fill in each ☐ with a number.

① $4 \times 7 = \boxed{} \times 4$

② $6 \times 8 = 8 \times \boxed{}$

③ $9 \times \boxed{} = 3 \times 9$

4 Let's find all the multiplication expressions that give the following answers.

① 4   ② 6

③ 18   ④ 24

5 The following diagram is a part of the multiplication table.

What are the appropriate numbers for ⓐ and ⓑ?

			ⓐ	
20	24	28	32	36
		ⓑ		

6 Let's think about how to calculate 12 × 4. Let's fill in each ☐ with a number.

(1) The answer to 12 × 4 is the same as the answer to ☐ × 12.

(2)
4 × 8 = ☐ ⎫
4 × 9 = ☐ ⎬ + ☐
4 × 10 = ☐ ⎬ + ☐
4 × 11 = ☐ ⎬ + ☐
4 × 12 = ☐ ⎭ + ☐

So, 12 × 4 = ☐

7 Let's think about how to calculate 11 × 2. Let's fill in each ☐ with a number.

(1) The multiplicand 11 can be decomposed into 6 and ☐.

(2) 6 × 2 = 12
☐ × 2 = ☐
12 + ☐ = ☐
So, 11 × 2 = ☐

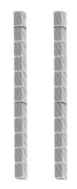

14 Fractions
pp.50~57

1 What is the fraction of the size of the colored part to the original size?

①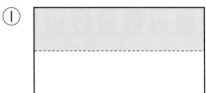

②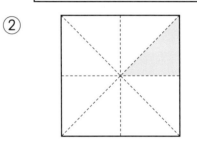

③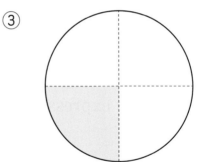

2 Let's color $\frac{1}{4}$ of the original size.

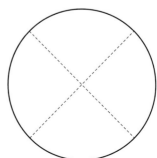

⑮ Time and Duration (2)
pp.60~65

1 Let's find the following time and duration.

① The duration from 8:00 a.m. to 10:00 a.m.

② The duration from 3:00 p.m. to 6:00 p.m.

③ The time 3 hours past 10:00 a.m.

④ The time 2 hours before 4:00 p.m.

⑤ The time 20 minutes past 9:10 p.m.

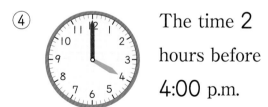

⑥ The time 15 minutes before 2:00 p.m.

⑯ Numbers up to 10000
pp.66~77

1 Let's write the following numbers.

① four thousand eight hundred sixty-two

② nine thousand three hundred seven

③ five thousand five

2 Let's write the following numbers.

① The number that is the sum of 3 sets of 1000, 7 sets of 100, and 4 ones.

② The number that is the sum of 6 sets of 1000 and 5 sets of 10.

③ The number that is the sum of 8 sets of 1000 and 9 ones.

④ The number that is the sum of 70 sets of 100 and 30 ones.

⑤ The number that is 1000 larger than 7000.

⑥ The number that is 50 smaller than 4000.

3 For the number 6700, let's fill in each ☐ with a number.

① 6 in the thousands place means that there are 6 sets of ☐.

② The number that is the sum of 6000 and ☐.

③ The number that is the sum of ☐ sets of 100.

④ The number that is 300 larger than 6700 is ☐.

4 Let's fill in each ☐ with a number.

① 3800 is the number that is the sum of ☐ sets of 100.

② 3800 is the number that is the sum of ☐ sets of 1000 and ☐ sets of 100.

③ 4290 is the number that is the sum of ☐ sets of 10.

④ 4290 is the number that is the sum of ☐ sets of 1000, ☐ sets of 100, and ☐ sets of 10.

5 Let's fill in the ☐ with > or <.

① 3800 ☐ 3790
② 8990 ☐ 9110
③ 5505 ☐ 5055
④ 7221 ☐ 7222

6 Let's answer by using the line of number below.

5000 ──────── 7000

ⓐ ⓑ ⓒ

① Let's write the numbers that are pointed by ↑ at ⓐ, ⓑ, and ⓒ.

② Let's write the number that is 400 larger than ⓐ.

③ Let's write the number that is 600 larger than ⓒ.

④ Let's write the number that is 400 smaller than ⓑ.

7 How many larger is 10000 than 9000?

⑰ Length (2)
pp.80~87

◪1 Let's explore the lengths of tapes.

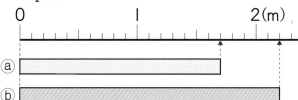

① What are the lengths of tapes ⓐ and ⓑ in m and cm?

② What are the lengths of tapes ⓐ and ⓑ in cm?

◪2 Let's fill in each ☐ with a unit.

① The length of a swimming pool ······25 ☐

② The thickness of a notebook ······3 ☐

③ The height of a cup ······10 ☐

◪3 Which is longer?

① 405 cm, 3 m 95 cm

② 1 m 10 cm, 108 cm

③ 2 m, 203 cm

④ Let's fill in each ☐ with a number.

① 3 m = ☐ cm

② 1 m 95 cm = ☐ cm

③ 500 cm = ☐ m

④ 208 cm = ☐ m ☐ cm

⑤ Let's calculate the following.

① 4 m 20 cm + 5 m

② 3 m 40 cm + 35 cm

③ 8 m 10 cm − 6 m

④ 7 m 65 cm − 2 m 20 cm

⑱ Addition and Subtraction
pp.90~105

◪1 There are 28 boys and 26 girls. How many children are there altogether?

① Let's fill in the ☐ with a word.

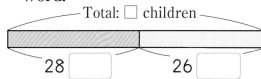

Total: ☐ children

28 ☐   26 ☐

② Let's write a math expression and the answer.

2　There were 40 cookies. 18 of them were eaten.

　　How many cookies are left?

① 　Let's fill in the ☐ with a word.

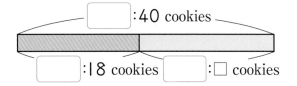

☐:40 cookies

☐:18 cookies　☐:□ cookies

② 　Let's write a math expression and the answer.

3　Yota has 34 sheets of colored paper. His brother has 8 more sheets than him.

　　How many sheets does his brother have?

① 　Let's fill in the ☐ with a number.

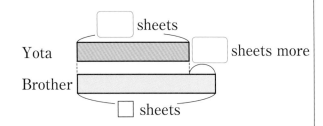

☐ sheets

Yota

☐ sheets more

Brother

□ sheets

② 　Let's write a math expression and the answer.

4　Akari has 35 marbles. Yuna has 7 fewer marbles than Akari.

　　How many marbles does Yuna have?

① 　Let's fill in the ☐ with a number.

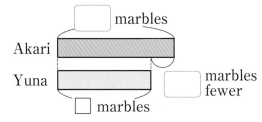

☐ marbles

Akari

Yuna

☐ marbles fewer

□ marbles

② 　Let's write a math expression and the answer.

5　There were 18 passengers on the bus. Since several people came on board later, there are 25 passengers now. How many people came on board?

① 　Let's write a math expression to find the total number of passengers.

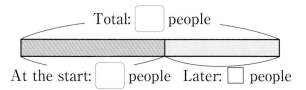

Total: ☐ people

At the start: ☐ people　Later: □ people

② 　Let's write a math expression to find the answer, and then write the answer.

6 At a cake shop, 28 cakes were sold and 14 cakes are left.

How many cakes were there at the start?

① Let's write a math expression to find the number of remaining cakes.

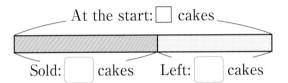

At the start: ☐ cakes

Sold: ☐ cakes   Left: ☐ cakes

② Let's write a math expression to find the answer, and then write the answer.

20 Shapes of Boxes

pp. 109~117

1 As for the shape of the box below, let's answer the following.

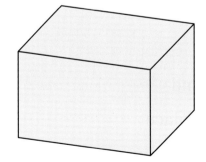

① How many faces does the shape have?

② How many edges does the shape have?

③ How many vertices does the shape have?

2 Let's fold the following figure of an opened box.

Which box is formed, ⓐ or ⓑ?

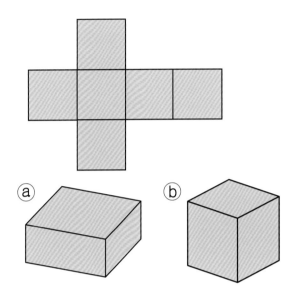

3 How many faces ⓐ, ⓑ, and ⓒ does the box shown below have respectively?

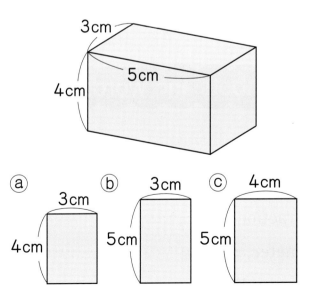

Symbols and words in this book

Multiplication Game ①

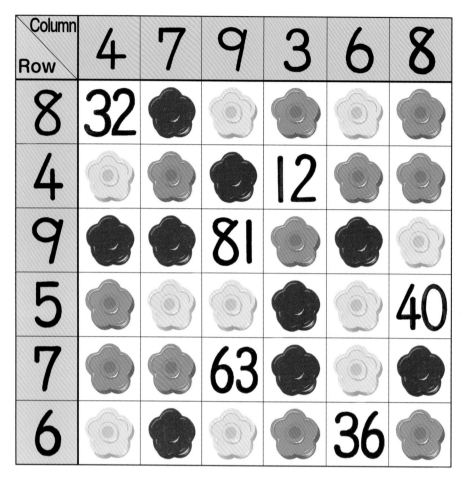

Dice for column

Dice for row

① Multiply the numbers in the row and the numbers in the column. Write the answers in the squares. Then cover the answers with marbles. Repeat this until you have used 30 marbles.

② Roll the 2 dice at the same time. Multiply the 2 numbers on the dice and say the answer. When the answer is correct, you can get a marble from the square.

③ If there is no marble on the square you choose, you have to put one of your marbles on it.

④ Decide the number of times to roll the dice and play in turns.

⑤ The person who gets the most marbles wins.

Multiplication Table

Multiplier

Multiplicand

×	1	2	3	4	5	6	7	8	9	10	11	12
1	1	2	3	4	5	6	7	8	9			
2	2	4	6	8	10	12	14	16	18			
3	3	6	9	12	15	18	21	24	27			
4	4	8	12	16	20	24	28	32	36			
5	5	10	15	20	25	30	35	40	45			
6	6	12	18	24	30	36	42	48	54			
7	7	14	21	28	35	42	49	56	63			
8	8	16	24	32	40	48	56	64	72			
9	9	18	27	36	45	54	63	72	81			
10												
11												
12												

10 × 3 is considered as
10 + 10 + 10.
So, the answer is 30.

Hiroto

When the multiplicands are 10 and
11, can I fill in the blanks of the
table with answers?

Yui

Multiplication Game ②

Enjoy with all your classmates.

① Fill in the table that has 16 squares with your favorite numbers from answers in the multiplication table.

② A person chooses a card from a complete set of 81 multiplication cards.

③ Find the answer and write a ○ if it is in the table you have made.

④ Repeat steps ②-③. A person gets 1 point when there is a ○ on every number in a row, column, or diagonal line.

⑤ Choose 40 cards in all. The person who gets the most points wins.

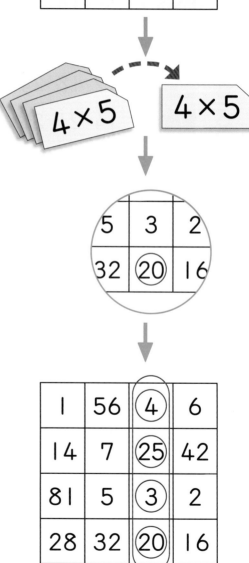

1 point

First Time

(empty 4×4 grid table)

Second Time

(empty 4×4 grid table)

Third Time

(empty 4×4 grid table)

Fourth Time

(empty 4×4 grid table)

Let's try a table with **25** squares.

What number should I write at the center of the table?

Yui

(empty 5×5 grid table)

Editorial for English Edition:

Study with Your Friends, Mathematics for Elementary School

2nd Grade, Vol.2, Gakko Tosho Co.,Ltd., Tokyo, Japan [2020]